I0002211

YOU DESERVE A TECH UNION

ETHAN MARCOTTE

Copyright © 2023 Ethan Marcotte
All rights reserved

Development editor: Lisa Maria Marquis
Editor: Caren Litherland
Expert reviewer: RV Dougherty
Book producer: Ron Bilodeau

ISBN: 979-8-9915420-0-5

www.heavychange.com

10 9 8 7 6 5 4 3 2 1

TABLE OF CONTENTS

For Grandma

FOREWORD

LABOR EXPLOITATION IS RAMPANT in the tech industry. We've watched it manifest in the rise of gig work orchestrated to atomize and exhaust workers. We've seen employee "perks" handed down to keep us working longer hours, rather than giving us fair pay. We've been subjected to mass layoffs that make management richer, all while disempowering workers and suppressing our wages.

As the industry matures, worker exploitation seeps deeper and deeper into the status quo — and a flourishing future once promised by venture capitalism violently breaks down. But so does one of tech's original cultural tenets: individualism. This collective awareness, this shared awakening helps us realize that our power comes not from climbing ever-narrowing ladders, but from linking arms and taking collective action. A realization that the only path forward is through change, now.

In *You Deserve a Tech Union*, Ethan captures a growing desire among tech workers to enact that change, sharing essential perspectives on what it means to truly embody solidarity and practically exercise collective action. He weaves together essential historical context and stories from experienced practitioners, offering a guide for creating and leveraging the kind of power found only among united workers.

As I read this book, I am reminded of a lasting truth: workers do not fight for power, we fight for dignity. Power is just a means to an end. If you're reading this now, you're ready to unite with your coworkers in fighting for — and claiming — the dignity you deserve.

— Clarissa Redwine

INTRODUCTION

MY GRANDMOTHER WORKED a dairy farm. She taught me so much while she was alive, more than I could possibly list here: about how to put in a day's work; about how to listen; and about the importance of making sure everyone's welcome at your table, and that they're well fed. And as I start this book, I can hear her voice reminding me to always say what you're going to say, as plain as you can.

So I'll start here, as plain as I can.

This is a book about power, and about work, and about the power we find in our work. It's also a book about history and the law, and the power we can find in those, too. This is a book about how the tech industry has changed over the past few years; it's also a book about what happened in the two centuries before that. It's a book about twenty thousand textile workers in Lawrence, Massachusetts striking in 1912; it's a book about twenty thousand Google workers walking out of their offices in 2018. It's a book about how the tech industry is changing, and about the people who are working to change it.

This is, above all else, a book about unions in the tech industry: what they are, why they matter, and how you and your coworkers can form your own. And by doing the work of forming a union, you're about to make your industry a little better than you found it.

Honestly, your timing couldn't be better. Because this is a damned weird time in the tech industry.

—okay, yeah, you're right. My grandmother would not have approved of that language. But a cuss or two is, sadly, warranted. Over the past decade or so, some hard questions have been asked about the tech industry's impact. The internet's early days were focused so much on the medium's promise, on how it could be used to connect people to the world's accumulated knowledge, and to each other. But in more recent years, it has become harder to square that vision against how the internet has been used to spread misinformation and disinformation, to destabilize governments, and to coordinate violence against marginalized and vulnerable people. And to top it all off, layoffs have wracked our industry in recent months, putting hundreds of thousands of people out of work.

Like I said. It's a weird, weird time.

But amid all that turmoil, something wonderful has started happening. In recent years, tech workers have realized that there is power in acting collectively — in identifying things they wish were different about their workplaces, and moving together to make those changes. And thanks to the tireless work of many, many people, there is a resurgent labor movement in our industry — a movement that's fighting for a better version of the tech industry.

That movement has taken many forms. Design teams have circulated open letters telling their management to correct discriminatory hiring practices. Employees have started sharing salaries with one another to address widespread pay disparities in their organizations. Tech workers have staged walkouts to advocate for safer workplaces free from discrimination, racism, and harassment. Others have taken to pushing back against their companies' military contracts, arguing that their employers should follow through on their stated corporate values.

But tech workers haven't stopped there. They have also formed unions. Over the past few years, digital teams at companies like Kickstarter and Glitch have unionized, as have the tech teams at NPR, the Atlantic, and the New York Times. Mission-driven organizations like Code for America, Change. org, and Nava are unionized, too. There are even unions at Alphabet and Apple — two upstart little tech companies you might have heard of. And those are just a few examples: many more tech companies have unionized, and many, many more are in the process of doing so.

That's why we're here, you and I — this is the story we're here to tell. In this little book, we're going to look at where our industry is right now. We'll talk a bit about the history of organized labor, both inside and outside the tech industry. We'll take a high-level look at how a union is formed, and I'll share some resources to help you and your coworkers get started. And then we'll talk about what might happen next.

Before we do, I should mention that this book's going to have a noticeable focus on the United States. I'll be writing about themes and issues that are truly universal, so regardless of the country you live in, I hope you'll find these pages valuable. But I do want to note that the histories I cover, and the legal contexts I mention, are both rooted in the country in which I live and write.

And I suppose there's one more thing I should make clear: I'm a bit of an odd choice to write about labor unions. I'm a working web designer, but I've worked for myself for much of my career. (Not all of it, but still.) I've been reading, writing, and speaking about labor issues in the tech industry for several years now, because it's a topic I care deeply about. But as a self-employed designer, I don't have much practical, hands-on experience with organizing a workplace, much less forming a union.

That's why I decided to talk to those who do have that experience.

Over the last year and a half, I've been speaking with the people who have been bringing unions back into the tech industry. I've interviewed tech workers whose unions have won their first contract, workers whose unions have just been recognized, and workers who are starting their first organizing campaign. I've met with full-time union organizers and staffers, who've graciously shared their strategies and lessons with me. I've spoken to activists, economists, and friends.

In other words, this book was shaped by many, many people. (The mistakes that remain, however, are mine alone.) I'm grateful for their time and their stories, and I hope that this book — in addition to everything else I hope it becomes — can act as a spotlight on their work.

And I hope that light will be useful to you. Because "forming a tech union" isn't some theoretical exercise. Amid all the pain and turbulence of the past few years, a labor movement exists in the tech industry right now, today. And that movement is focused on building protections, safety, and power for tech workers — for you, and for me.

Sound good? Marvelous. Because we've got some good work ahead of us, you and I. Let's get started.

1 JUST WORK

> "I remember my mother said life can be better — you have to help make it so."
>
> —From an interview with Rev. Addie L. Wyatt (https://ydatu.com/01-01)

WE DON'T REALLY TALK ABOUT power in the tech industry. And I've always thought that was a little odd.

I mean, we often talk *broadly* about power. We talk about "the power of design" to reach people; "the power of the internet" to connect people; "the power of platforms" to solve problems. But those phrases feel a little abstract, don't they? By and large, the tech industry as a whole doesn't spend much time discussing the shape or impact of our industry's power. Seems...weird, right?

I suppose it'd be helpful to *define* "power." So let me ask you: Have you come across Ursula Franklin's work before? If not — well, I'm not sure where to start. I mean, she was a metallurgist, an experimental physicist, a university professor, a feminist, a social historian...honestly, I could keep going. She's easily one of the most inspiring, most accomplished people I know of. In addition to everything else, Franklin was the author of a book called *The Real World of Technology*. And in that book, Franklin talked extensively about the power that technology has — not as an abstract concept, but as a social and political force. One that has the power to shape, and reshape, people's lives.

When it comes to defining power, Franklin has an approach I've always liked. She talks about *plans*:

> Webster defines [planning] as "making plans ... arranging beforehand." I like this simple definition because it says that there are planners as well as plannees; there are those who plan and those who conform to what was arranged before-hand. Just as it is easier to give good advice than it is to accept it, it is much more fun to plan than it is to be subjected to plans made by others. (https://ydatu.com/01-02)

Immediately, something comes into shape here: there are those who *make* plans, and those who are *subjected* to the plans. This feels like a pretty good framework for defining "power," because there's an explicit hierarchy: one of these roles makes decisions; those decisions then shape the behavior of others, who must conform to the plans.

THE WORK OF THE WEB

When Franklin talks about plans, she's talking in part about how they operate in the context of *work*. After all, work has plenty of *planners* — the people who make the plans — and it has plenty of *plannees* who are subjected to those plans. And regardless of the type of work you do, I bet you probably oscillate between both ends of that spectrum throughout your day, or on different projects. Maybe you're the design lead on a new contract at your agency job, and you're in more of a *planning* role; or maybe you're an engineer implementing a new product, acting a bit more like a *plannee* while you follow someone else's spec. But once again, that matter of hierarchy rears its head: there's power involved in creating a plan, as it directly influences the behavior of those who follow it.

And I'd bet good money that *your* work has been affected by someone else's decisions further up in your company. Maybe your boss decided that you had to start taking on additional responsibilities, and you had to figure out a way to start juggling them against your existing workload. Or maybe you've spent months working on a new project, only to have your company's leadership suddenly end its funding, and then move you onto a different team. It's even possible that you lost your job this year, due to decisions made by your company's leadership. (And if that happened to you, I'm truly sorry.)

If any of this sounds familiar to you, I'd like to suggest that your work is not primarily about creating plans; instead, your work is shaped and defined by the plans other people make. It's true that you contribute technology-based labor to your employer to support their plans and business goals; in exchange, you're paid for the work you do. Ultimately, though, you're not the one creating the plans — by and large, you're expected to follow those plans.

Now, your job could involve engineering, design, content, research, customer support, moderation — or some combination of the above. (Or something else entirely!) But regardless of the kind of work you do, this two-tier power dynamic still applies. If your labor contributes to a digital product or service, and you are compensated for that labor, then you perform *tech work*. You are a *tech worker*.

You may not work in the technology industry proper, mind. I mean, sure: maybe you *do* work for a massive tech company, for an early-stage startup, or somewhere in between. But every sector of every industry has some form of tech work — it's just as likely that you do tech work in healthcare, finance, education, or in some other field altogether. In other words, tech workers exist in every industry. If your job is technical in nature, and has ever been defined — or redefined, or eliminated — by the decisions made by your company's management or leadership, then you are a tech worker.

Why does this matter? Well, now that we've roughed out a working definition of power, and looked at examples of how that power can manifest itself in our jobs, we're ready to take a look at how that power impacts *us* — the people whose work built this industry.

THE TECH WORK YOU DO

If you'll indulge me, I have two questions for you:

1. What do you like about your job?
2. What would you change about your job, if you could?

You don't need to answer these questions right away. But maybe take a minute or two at some point soon, and think about them: If you had to list what you liked — or didn't like — about your job, what would you write down?

I bet there's a lot you like about your job. For example, it's possible you're happy with your salary. After all, it's not uncommon for folks to hear the phrase "tech worker," and then think the phrase "highly paid." It's not hard to see why: in 2021, the median income in the United States was $69,717 (https://ydatu. com/01-03); but that same year, the median salaries at Alphabet and Meta were estimated to be almost $300,000 (https://ydatu. com/01-04, subscription required).

That is an eye-watering number, to be sure. But I'll note that's a *median* salary — not every worker at Alphabet and Meta makes that amount of money. Moreover, not every tech worker is employed by a company as big (or as affluent) as Alphabet or Meta: maybe you work at a much smaller startup, or at a small agency that designs websites for small businesses. With all that said, it is worth acknowledging that, *on average,* tech workers enjoy a considerable amount of privilege relative to other kinds of work: we work indoors, out of the elements; our pay tends to be pretty good, as do our benefits.

That's why listing what you like about your job can be a useful exercise. But it's important to answer the second question, too: What would you *change* about your job? Because every job has benefits, sure — but it also has drawbacks. Maybe you're pretty happy with your salary, but you wish you had better health insurance. Or maybe you're satisfied with the number of vacation days you get each year, but you feel like you're paid less than your peers, or that you're overdue for a promotion. Maybe your employer refuses to hire more staff, but your workload keeps increasing. (Along with your hours, and your stress.)

Maybe you're concerned about layoffs, and worried you won't get any severance pay if you're forced to leave your job. Maybe you've been harassed or discriminated against at work, and you don't think that's even remotely acceptable.

Now, these are just a few answers I cooked up; your answers might be dramatically different. With that in mind, here are two more questions for you:

1. For the things you like about your job: *How would you keep them from changing?*
2. For things you wish were different about your job: *How would you change them?*

These questions might feel a bit trickier to answer.

For the first question, there's very little assurance that the benefits you enjoy today might be there tomorrow. Even in the past year, we've seen tech companies slashing not just staff, but benefits. News reports tend to focus on cuts in extravagant-sounding perks at the largest tech companies, like when Meta announced in early 2022 that it was ending free laundry service for employees, and cutting back on cafeteria hours (https://ydatu.com/01-05). But while the benefits you have at your job might not be that indulgent, I bet they matter to you. Maybe your company's modest training budget just got axed. Maybe your employer's health insurance plan changed, and some critical coverage you depend on won't be available next year. Or maybe your employer announced an end to your company's remote work policy, but you don't feel it's safe to return to the office amid the ongoing pandemic.

The second question — *How would you change them?* — might be more difficult to answer. Depending on *what* you want to change, your path to making that change might feel scary, if not downright impossible. Advocating for yourself to get a raise might feel doable if you have a good relationship with your manager. But what happens when that manager leaves? How do you fight for a path to promotion when you *don't* have a manager who's invested in your career growth?

THE POWER IS ALREADY HERE, IT'S JUST NOT EVENLY DISTRIBUTED

I'd like to suggest that there's a considerable amount of *uncertainty* and *unpredictability* inherent in our relationship to tech work—that our role as workers is defined almost entirely by a lack of guarantees. The benefits you enjoy aren't guaranteed to you; they're offered entirely at your employers' discretion, and might change in a year's time. (Or in a month's time.) Your path toward a raise or promotion might depend less on a fair and transparent advancement system, and more on whatever social capital you've managed to accrue at work. You might find yourself reassigned to a new team, doing work that's completely different than what you interviewed for. Your job might exist tomorrow—or it might not.

And that's never felt truer than it does right now. Because the last twelve months in the tech industry have been, frankly, tumultuous. The global pandemic created an unprecedented demand for online goods and services, which dramatically buoyed the tech industry's profits. By the end of 2021, the industry was booming; it literally couldn't hire quickly enough. Tech companies were opening new job positions faster than nearly any other industry, competing aggressively on salary, benefits, and other perks to attract qualified candidates. Heck, Amazon even doubled its salary caps at the start of 2022, raising the annual base pay for salaried workers from $160,000 to $350,000 (https://ydatu.com/01-06, subscription required).

By the end of 2022, things could not have felt more different. In February, a land war began in Europe, causing countless deaths and sparking a widespread humanitarian crisis. But in meaner terms, that tragedy also upended global markets and supply chains, causing inflation and costs of living to skyrocket. And the tech industry responded in kind. In some cases, corporate leaders issued notes of caution, with a hint of foreboding mixed in. Citing the global downturn, Google's CEO announced in mid-2022 that the company was slowing hiring; he also insisted his remaining staff should work with "greater urgency, sharper focus, and more hunger" (https://ydatu.com/01-07).

But since then, mass layoffs have shaken our industry. According to the website Layoffs.fyi, an open tracker of layoffs in the tech industry, more than 160,000 workers in the tech industry lost their jobs in 2022 (https://ydatu.com/01-08). Here are a few notable examples:

- Four months after raising its salary caps to attract new candidates, Amazon laid off ten thousand corporate and technology jobs at the end of 2022 — and is reportedly considering plans to lay off ten thousand more (https://ydatu.com/01-09).
- After a costly pivot into virtual reality, Meta has seen its once-indomitable user base tumble sharply alongside its stock price; as a result, Meta's leadership laid off eleven thousand employees at the end of November 2022 — and then, four months later, it laid off another ten thousand (https://ydatu.com/01-10, https://ydatu.com/01-11, subscription required).
- In January 2023, Microsoft announced it would be reducing its global workforce by around five percent, a move that would cause ten thousand people to lose their jobs (https://ydatu.com/01-12).
- Two days after Microsoft's announcement, Google announced that twelve thousand workers would be laid off (https://ydatu.com/01-13, subscription required).

It's important to note that these aren't equal-opportunity layoffs: they have done substantial harm to the workers who were already underrepresented in our industry. Over the last few years, some progress has been made in improving the tech industry's longstanding issues with diversity. But in 2023, Stack Overflow published data showing that those modest gains haven't just been erased by this torrent of layoffs — they're actively being reversed:

Of those who lost their jobs in the most recent round of layoffs, 45% were women — which doesn't sound bad until you remember that less than a third of tech industry roles and less than a quarter of tech leadership roles are filled by women. Other underrepresented groups, especially Black tech workers, have also been impacted at outsize rates. [...]

Women and people of color aren't being laid off at higher rates because we were dead-weight DEI hires in the first place. According to Sarah Kaplan, director of the University of Toronto's Institute for Gender and the Economy, it's because "the roles that historically underrepresented groups are hired into tend to be seen as the most expendable." (https://ydatu.com/01-14)

As with the shift to remote work, these layoffs are exacerbating existing harms, and visiting those harms on the industry's most vulnerable workers.

Of course, being suddenly forced out of your job is a clear example of the dramatic power differential in play between you and your employer. But even in times less dominated by layoffs, that unequal power distribution still exists. When you started your job, you agreed to your salary, and to the benefits that were offered. Maybe you even negotiated for a better salary, or a better title, or both. But beyond that, you agreed to what was handed to you, and didn't have much recourse to changing the terms of your employment.

Outside of leaving your job, that is. I spoke anonymously to one tech worker involved with a union organizing effort at a large media company. One of the drivers behind their union campaign was the feeling that the company didn't provide any opportunities for career growth, which caused a ridiculous amount of turnover among the staff. "Often, the only way tech workers could advance their career was by applying elsewhere," they told me.

Now, that's a perfectly valid response, especially to a toxic or unsafe work situation. But of course, a new job doesn't actually guarantee that things *will* be substantively better. The top-line numbers might improve — a better salary, more perks, a more flexible schedule — but the same lack of certainty remains. What's more, the organizer I spoke with mentioned something interesting: "voting with your feet" has dramatically diminishing returns not just for individual workers, but for everyone. They told me:

Having this large pool of candidates made it so that when workers asked for improvements about their working conditions, or wanting to change things, management often puts on this real, like, "take it or leave it" stance: "this is what we're offering."

This struck me as a rather profound point. By and large, workers in the tech industry have come to expect that a bad job can only be fixed by getting a better job somewhere else. (And that there *will* be another job. As 2022 has shown us, that isn't necessarily the case anymore.) But this system of "fixing it by moving on" affects employers, too: it has trained them to expect that *there will always be another worker.* If one employee doesn't care for their healthcare plan, for their compensation, for having to work weekends, for their toxic manager? Well, that's too bad: *this is what we're offering.* If you don't like it, you're more than welcome to leave.

Not everyone is able to leave as easily as that, even in the best of times. And as I write these words, these are assuredly not the best of times — in the middle of a tight labor market, in which every other headline seems to contain news of more industry layoffs. But even still, there are layers of privilege that influence whether or not you can leave a job. If you're an engineer, it's possible that you've historically enjoyed more demand than your peers in design or research, or your coworkers who run the front desk at your office. But if you have coworkers whose salaries aren't enough to cover their living expenses, they might be holding down a second job to make up the difference. It's harder to conduct a job search when your nights and weekends aren't free.

Just to draw a line under it: while tech workers tend to enjoy a considerable amount of privilege relative to other industries, that privilege isn't distributed equally across all tech workers. As one example, let's look at how many of our employers have embraced a shift toward remote work, driven by the COVID-19 pandemic. Project Include, a nonprofit that studies issues of diversity and inclusion in the tech industry, published the

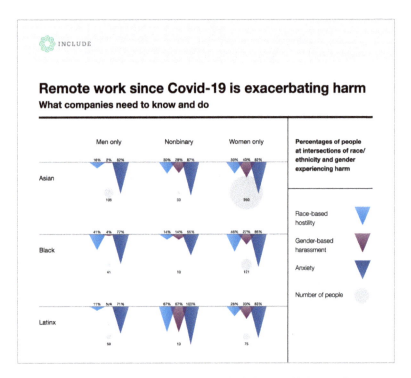

Remote work since Covid-19 is exacerbating harm
What companies need to know and do

FIG 1.1: A chart from Project Include's report on the industry-wide harms of remote work, showing how distributed work has led to an increase in distributed harassment and discrimination.

results of research they'd conducted on how that shift had impacted tech workers (https://ydatu.com/01-15).

Generally, those impacts aren't great. It's true that remote-first offices opened tech workers up to considerably more geographic freedom, and insulated us from the worst effects of the global pandemic. But Project Include found significant drawbacks, too: most of their respondents said they experienced increased anxiety and worked longer hours. More than a third felt increased pressure to be online and available for their managers — even outside of normal working hours.

Additionally, Project Include found that the shift to remote work *compounded* old forms of workplace discrimination and harassment (FIG 1.1). A quarter of the respondents said they'd encountered an increase in gender-based harassment; one in *ten* experienced an increase in race- or ethnicity-based hostility. And this hostility took many forms, thanks to the productivity tools favored by remote workplaces — ranging from bullying on group video calls or emails, to racist and sexist links shared in chat software.

The scale of the harms is new, but the harms themselves are not. The tech industry has long been predominantly white and male-dominated. It is also rife with gender- and race-based disparities in pay: according to a 2021 survey, companies offer women 2.5 percent less on average than men; Black and Hispanic women earned $0.90 for every dollar paid to their white male coworkers (https://ydatu.com/01-16).

But the shift to remote work has brought new harms to its workforce — and because of the industry's preexisting biases, those harms aren't distributed equally. Here's Project Include again (https://ydatu.com/01-15):

> The pandemic-driven shift to remote workplaces has exacerbated longstanding, systemic problems and amplified workplace biases. Bad management and communications got worse, as did anxiety and work-life balance, especially for people from marginalized communities. (https://ydatu.com/01-15)

So. We're left with a relationship with our jobs that's mired in uncertainty, unpredictability, and a lack of guarantees. It's a relationship defined by a wildly uneven distribution of power between employers and the people who work for them. And as ever, certain tech workers are exposed to considerably more bias and harm in the workplace than others.

On top of all of that, the job you have today might not be the job you have tomorrow. And that suits your bosses just fine.

WORK (PROBABLY) WON'T LOVE YOU BACK

After months of legal drama, the billionaire Elon Musk finally purchased Twitter at the end of 2022. After completing the sale, he began gutting the social media company, firing approximately half of his employees—roughly 3,700 salaried workers—*over email*. Entire teams at Twitter were hollowed out, with some learning they'd been dismissed in the middle of the night. Some people were "accidentally" fired, and then asked to return so they could continue working on their revenue-generating products. And then some of them were fired again (https://ydatu.com/01-17).

Having completed the mass firings, Musk emailed the remaining staff to inform them they'd "need to be extremely hardcore," and to commit to working "long hours at high intensity" (https://ydatu.com/01-17). Subsequently, reports emerged that employees were sleeping in Twitter's offices—first in sleeping bags, then in conference rooms that'd been converted to bedrooms (FIG 1.2).

The uncertainty, the humiliation, the expectation that you effectively live in your office, the verbal abuse cast about by a so-called leader of your company—no person should have to go through an experience like the workers gracelessly fired at Twitter. A lot of this comes from the fact that the tech industry *overwhelmingly* hires workers on an "at-will" basis, which allows an employer to fire an employee at any time, for any reason. (Or for no reason at all.) And Musk exploited this to terrible effect.

The situation at Twitter makes quite a contrast to Stripe, which also conducted a round of layoffs in November 2022. Here's the email that Stripe's CEO sent out to his employees:

> Around 14% of people at Stripe will be leaving the company. We, the founders, made this decision. We overhired for the world we're in (more on that below), and it pains us to be unable to deliver the experience that we hoped that those impacted would have at Stripe.

FIG 1.2: The escalation of Twitter's "extremely hardcore" work culture: a tweet from Esther Crawford, a director of product development, showing her sleeping during a crunch (https://ydatu.com/01-18), registration required); roughly a month later, conference room sofas were covered with bedding (https://ydatu.com/01-19).

> There's no good way to do a layoff, but we're going to do our best to treat everyone leaving as respectfully as possible and to do whatever we can to help. (https://ydatu.com/01-20)

Tonally, this is worlds better than anything that came out of Twitter. But more importantly, the CEO proceeds to spell out the benefits that impacted employees would receive. And they're quite generous: Stripe committed to paying out fourteen weeks of severance, as well as all unused PTO time; they also agreed to provide six months of healthcare coverage. They offered career support for all affected employees, offering to try to connect them with other companies that were hiring. And perhaps most crucially to my eyes, they offered immigration support to visa holders, recognizing that they were perhaps most impacted by layoffs in a tight labor market.

The layoffs at Twitter and at Stripe are, to be clear, two extremes: one is a case of cruel mismanagement executed by a billionaire whose deficiencies as a leader seem to run as deep as his contempt for his employees; the other is an example of layoffs conducted with respect for the people affected, and with an attempt to provide them with some financial security.

But I want to point out a common thread between these two firings: the nature of these layoffs was singly and solely determined by each company's leadership. In other words, the workers affected by these layoffs didn't have any say in how much notice they were given, or in the benefits and support they'd receive after leaving. That power sat exclusively with the two companies' respective bosses.

Stripe treated its workers well, it's true. But those benefits might not be so generous in a future round of layoffs. And besides, shouldn't those benefits be the norm? Wouldn't you like to have a *guarantee* that, if you lose your job, you'll receive that same level of care? Heck, I'll do you one better: wouldn't you like to have a hand in defining the benefits you receive at work? After all, our labor—yours and mine—built this industry. We *should* have a seat at that table. But at the moment, we don't.

Let's talk about how we'll change that.

2

IN OUR STRENGTH, SAFETY

"Economic injustice doesn't happen by accident."
—David Delmar Sentíes, *What We Build with Power* (https://ydatu.com/02-01)

TO GAIN POWER FOR OURSELVES as tech workers, we have to take a look at where we work. Maybe just as importantly, we have to talk about how our industry wields its power — both for us and against us. And that means having a frank conversation about what our work is used *for*.

I'll start here, with something I read a few years ago.

In 2018, United Nations human rights investigators issued a report stating that Myanmar's military targeted its population of Rohingya Muslims with "genocidal intent." More specifically, the Rohingya population was subjected to systematic, planned violence, including mass killings and gang rapes. And in their report, the UN investigators said that Facebook had been a key facilitator of that genocide. Here's the key paragraph:

> The role of social media is significant. Facebook has been a useful instrument for those seeking to spread hate, in a context where, for most users, Facebook is the Internet. Although improved in recent months, the response of Face-

book has been slow and ineffective. The extent to which Facebook posts and messages have led to real-world discrimination and violence must be independently and thoroughly examined. (https://ydatu.com/02-02, PDF)

"Facebook has been a useful instrument for those seeking to spread hate."
I first read that line years ago, but it still haunts me — as does the rest of the report. We've known for quite some time that Facebook, Meta's flagship product, has been used to facilitate real-world political attacks. Video footage of mass shootings has frequently appeared on the platform, and the company has scrambled to take it down. The platform is no stranger to misinformation and disinformation, either: over the years, conspiracy theories have run rampant across Facebook, leading to real-world attacks and violence.

But when I read this UN report, it felt new. Terrible, and new. Here was a government using a social media platform to help it visit unspeakable horrors on a vulnerable and marginalized population. This social media platform — one of our industry's most prominent products, in fact — was found to be a useful instrument to spread hate by a country's military force.

THE INDUSTRY THE WEB BECAME

Facebook isn't the only technology company complicit in state-sanctioned violence. The most prominent companies in our industry have become, in a word, militarized:

- In 2018, Amazon developed its own facial recognition software called Rekognition, and began selling it to law enforcement agencies in the United States (https://ydatu.com/02-03). In the years since, Amazon's investments in video surveillance and facial recognition have only deepened, along with its ties to law enforcement and intelligence agencies. In 2021, Amazon's Ring cameras were reported to have "connected around one in ten police departments across the US with the ability to access recorded content

FIG 2.1: Four soldiers wearing Microsoft's high-tech Integrated Visual Augmentation System (IVAS) goggles (https://ydatu.com/02-07).

from millions of privately owned home security cameras" (https://ydatu.com/02-04).

- The United States Department of Defense recently announced a partnership with Google, "setting the stage for potential broader adoption [of Google Cloud] across defense agencies worldwide" (https://ydatu.com/02-05).
- Not to be left behind by its competitors, Microsoft has also gotten into the government contracting game. Whereas Alphabet's early forays were almost exclusively software-based, Microsoft has been selling software and hardware to the United States military. In fact, they've created something called an Integrated Visual Augmentation System, or IVAS. These high-tech goggles are designed to provide a real-time, heads-up display when worn by soldiers in the field (FIG 2.1). Reportedly, the contract could be worth nearly $22 billion dollars over the next decade (https://ydatu.com/02-06, subscription required).

When I look at these developments, it's hard not to feel a bit of cognitive dissonance—and more than a little despair. Because as someone who effectively grew up alongside the tech industry, it's honestly difficult to reconcile the idealism of *the web I was promised* with *the web we got*. In the early days of my career, it was hard not to stumble over rhetoric about how the internet was going to better humanity.

How did we get here?

HOW TECHNOLOGY AGES

In Chapter 1, I mentioned Ursula Franklin's *The Real World of Technology*. I wanted to start with Franklin's book not just because it was incredibly prescient, or because I love reading it. (Though it was, and I very much do.) But Franklin also helped me understand *how* the tech industry ended up at this point in its history—this arguably low point. You see, Franklin suggests that as a given technology matures, it passes through several stages, each of them defined by that technology's relationship to society as a whole. More importantly, she argues that *every* technology follows this trajectory.

Here's one example Franklin cites in her book. In the middle of the nineteenth century, the sewing machine was introduced to the general public for the very first time. How was it introduced, you ask? Well, here's an excerpt from an article written in 1860, describing the benefits of this wondrous new machine to its readers:

> The sewing machine will, after some time, effectively banish ragged and unclad humanity from every class. In all benevolent institutions, these machines are now in operation, and do—or may do—100 times more towards clothing the indigent and feeble than the united fingers of all the charitable and willing ladies collected through the civilized world could possibly perform. (https://ydatu.com/02-08)

What's happening here? Well, when the sewing machine was first made commercially available, it was advertised as an appliance that would free women from the routine drudgery of hand-sewing. But that's not all: the authors of this article claimed that the sewing machine would "banish ragged and unclad humanity." Put another way, the authors were promising that *the sewing machine would end poverty.*

Now, I realize that all sounds a bit hyperbolic. (And, well, it is. This article was written in 1860; you and I both know the sewing machine didn't exactly fulfill that promise.) But Franklin notes that this is emblematic of how every new technology is introduced to society. Early adopters will enthusiastically extol the virtues of this bright, shiny new thing, telling stories of all the solutions it will provide. What's more, the technology is usually presented as a kind of liberation: as a revolutionary innovation, one that will free its users from toil and drudgery, and offer a clear, compelling solution to life's woes and cares. (You can probably see where this is going.)

Hold on to that hyperbole for a second, because we need to jump forward a few decades and look at another piece of writing about the sewing machine. Here's an excerpt from a Singer instruction manual published in 1910 (FIG 2.2):

> This machine, for trimming and overedging knit goods, can be used at the highest practicable speed for all seaming, hemming, putting on cuffs, sewing on borders, edging armholes, necks, collarettes, and bottoms of garments, making a smooth, firm and sightly seam that is entirely finished when it leaves the machine.

Bit of a shift from the first article, isn't it? The lofty rhetoric is gone, replaced with writing that almost feels mechanical — even clinical. Now, maybe that's not surprising; this is an instruction manual, after all. But you might have noticed that the flowery prose isn't the only thing missing. There's nothing in this excerpt — or anywhere else in the manual — about ending poverty, or clothing the poor. Instead, there's an emphasis on technical efficiency: on how this machine enables its operator *to do more work, more quickly.* In fact, there's an entire section titled "Speed." (Bet you can guess what it covers.)

Okay, so: What just happened? Why did the language around sewing machines change so dramatically between these two

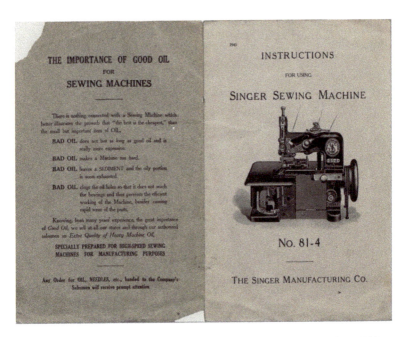

FIG 2.2: The first two interior pages of a Singer sewing machine manual from 1910 (https://ydatu.com/02-09, PDF).

examples? Inside of five short decades, the sewing machine is no longer discussed as a technology that will liberate its users. Instead, it delivers *productivity*.

According to Franklin, this is to be expected. Specifically, she argues that when every technology — *every* technology — is first introduced to a society, it's always heralded with enthusiasm, and with grand promises. But in effect, all that advocacy simply lays the groundwork for greater adoption. And as that technology continues to become more popular, it's no longer a novelty — instead, it becomes expected. A necessity.

It certainly happened to the sewing machine. That user manual was written at the turn of the twentieth century, when clothing was being manufactured at a truly industrial scale for the very first time. And the work was done by the people the sewing machine had initially promised to clothe: by poor,

FIG 2.3: A garment shop filled with women, sitting shoulder to shoulder at long tables lined with sewing machines. Photograph courtesy of flickr user Kheel Center under an Attribution 2.0 Generic Creative Commons license (https://ydatu.com/02-10/).

marginalized people, most of them women and immigrants. They would labor for long hours in factories and in sweatshops for very little pay, working in dangerous — and often deadly — conditions (FIG 2.3).

As it always does, that level of industrialization required a classic division of labor, which separated workers by tasks: one seamstress would be responsible for sewing up sleeves, another worker sewed the sleeves to the shirt, another worker would cut buttonholes, another worker would press the finished shirts. And then, as the industry evolved, those individual tasks — the sewing, the cutting, the pressing — became increasingly automated, excluding people altogether. Taken in total, this means that by the early twentieth century, the sewing machine was no longer a novel technology, promising to cure humanity's ills. Instead, it had become *institutionalized*.

But here's the critical point Franklin makes: as a technology ages, the initial promise of liberation is never, ever fulfilled. In other words, once a technology achieves institutionalized status, it reshapes people's labor — and people's lives — on a

dramatic, industrial scale. A technology may promise liberation at first—but inevitably, people are captured by it.
Exploited.

THE INTERNET, INSTITUTIONALIZED

Franklin's model feels almost eerily accurate. It's an endlessly
useful model, and it applies to just about any technology you
can think of: cars, personal computers, microwave ovens, your
favorite JavaScript framework, you name it. Each of them rode
a wave of enthusiasm to widespread adoption, and eventually
they become something akin to infrastructure.

But Franklin's framework is useful for me here, now, as I take
a hard look at the industry I grew up in. Because as I reread
some of these headlines—about one-time internet booksellers
selling facial recognition software to law enforcement agencies;
about giant software companies manufacturing augmented
reality headsets for soldiers—my mind immediately flashes
back to a different piece of writing altogether.

> We envision a world where all people are empowered by the
> Web. Everyone—regardless of language, ability, location,
> gender, age or income—will be able to communicate and
> collaborate, create valued content, and access the informa-
> tion that they need to improve their lives and communities.
>
> The creativity of the billions of new Web users will be
> unleashed. The Web's capabilities will multiply, and play
> an increasingly vital role in reducing poverty and conflict,
> improving healthcare and education, reversing global warm-
> ing, spreading good governance and addressing all chal-
> lenges, local and global. (https://ydatu.com/02-11)

This is the vision statement of the World Wide Web, as defined
by the Web Foundation (https://ydatu.com/02-12)—the nonprofit organization founded by Sir Tim Berners-Lee, the creator
of the World Wide Web. And it perfectly encapsulates the ideals,
the *optimism* that surrounded the early days of the medium.
There was a broad sense that this medium wasn't just new, but
revolutionary. After all, the web was a free, open network that

could be accessed by anyone, from anywhere in the world, by any internet-connected device — how could it *not* change everything for the better?

Well, here we are.

The sewing machine was heralded as a device that would end poverty and clothe the world's poor — but inside of fifty years, it had created a new class of industrial labor and impoverished the people it had promised to liberate. Meanwhile, the web promised to connect humanity, to reduce both poverty and armed conflict, and to reverse global warming; but today, the most prominent companies in our industry have become, at least in part, defense contractors.

Now, you might have noticed that my own morals and biases are showing a bit. On a personal level, I'm deeply troubled by the tech industry's entrenchment in the military-industrial complex — in the parts of our government that orchestrate war and violence. But having said that, I do understand if you feel differently. I respect that disagreement. Truly.

But regardless of how you and I respectively feel about the *morality* of this shift in the industry, I want to suggest that *the shift itself* is significant. We've come a long way from the utopian promise of an internet that will connect people across borders. Instead, we have an internet used to violently police those borders, and to inflict violence on vulnerable people. This is a clear, strong signal we've reached a new era in the tech industry's evolution: any industry participating in government contracting is no longer a novelty, much less a utopian one. It is, however, an industry that possesses an incredible amount of power.

Here's the thing, though: you and I have power, too.

WITH CONVICTIONS AND A VOICE

In 2017, a new American president came to power. Shortly thereafter, he instructed federal immigration authorities to begin forcibly separating families at the border — by quite literally taking children away from their parents and putting them in detention centers.

A few months before this new president was sworn into office, some fifty tech workers had collaboratively written the Never Again pledge, and posted it online. The first paragraph reads:

> We, the undersigned, are employees of tech organizations and companies based in the United States. We are engineers, designers, business executives, and others whose jobs include managing or processing data about people. We are choosing to stand in solidarity with Muslim Americans, immigrants, and all people whose lives and livelihoods are threatened by the incoming administration's proposed data collection policies. We refuse to build a database of people based on their Constitutionally-protected religious beliefs. We refuse to facilitate mass deportations of people the government believes to be undesirable. (https://ydatu. com/02-13)

In the week following its publication, over 2,500 people signed the pledge, demonstrating a wildly heartening degree of support for the pledge and its goals. It was signed by workers, managers, and executives from across the tech industry—from employees at companies as large as Google and Microsoft, to tech workers at government agencies, to self-employed designers and developers. In fact, the level of support was so great that it quickly overwhelmed the organizers, who had to stop accepting new signatures. They simply couldn't verify them fast enough.

The Never Again pledge is a remarkable piece of activism, one that shows the tech industry is *filled* with people who care about the industry's direction and ethics. Each and every one of those names represents a tech worker who's willing to ensure their labor isn't used for mass deportations, or for other forms of violent surveillance.

Keep that pledge in your mind for a moment, because I need to jump ahead two years, and talk about Microsoft. Now, I realize we've already discussed some of Microsoft's defense contracting work. But in 2019, one of the company's most high-profile controversies occurred when GitHub, a Microsoft-owned subsidiary, entered into a contract with Immigration and Customs Enforcement (ICE).

If the name's new to you, ICE is a sprawling federal law enforcement agency tasked with enforcing immigration law and detaining those seen in violation of it. And during the Trump administration, ICE was responsible for violently separating migrant children from their families, and then housing the children in facilities that were described as "prison-like." Reports abounded in the camps of overcrowding, inadequate care, and abuse (https://ydatu.com/02-14).

When GitHub's contract with ICE came to light, then-CEO Nat Friedman said that while it was "not financially material" to the company, GitHub would not be canceling the contract (https://ydatu.com/02-15). In response, employees at GitHub wrote a scathing open letter to Friedman, asking him to immediately cancel the contract. They said:

> We are not satisfied with GitHub's now-public stance on this issue. GitHub has held a "seat at the table" for over 2 years, as these illegal and dehumanizing policies have escalated, with little to show for it. Continuing to hold this contract does not improve our bargaining power with ICE. All it does is make us complicit in their widespread human rights abuses.
>
> We cannot offset human lives with money. There is no donation that can offset the harm that ICE is perpetrating with the help of our labor. (https://ydatu.com/02-16)

Once it was published, the letter received over 150 signatures — roughly a tenth of GitHub's estimated 1,300 employees — in less than an hour.

I mention this letter because it's a remarkable example of *collective action*, in which a group of people band together to address an issue. Here, GitHub's employees are approaching their company's top executive with a demand — and the demand has weight *because* they're acting collectively. A lone worker lodging a complaint could be ignored, or disciplined. (Or worse, dismissed outright.) But when a group of workers act together, it's much, much harder to disregard them.

When I reread their letter, I can't help but think of the Never Again pledge. In both cases, these letters are explicit acknowledgements that technology can be—and in fact is—used for harm. But there's something else in the GitHub letter, too: workers are demanding they should have a say in how their work is used. "There is no donation that can offset the harm that ICE is perpetrating with the help of our labor." It's right there in the text: *our labor*. We made this software, which is now being used to facilitate human rights abuses. And we're not going to stand for it.

WHEN TECH WORKERS WALK OUT

In the intervening years, workers have continued to hold their employers to account—or more specifically, to demand that their employers conduct themselves more ethically. In fact, over the past few years there has been a staggering amount of criticism of the tech industry's moral lapses, much of it led by workers inside the industry. Tech workers have been protesting decisions made by their employers; tech workers have been holding their leaders to account for abuses of power; tech workers are demanding safer working environments, free from harm and abuse; tech workers are walking out of their jobs to fight for something better.

Take Amazon, for example. As it happens, the massive online retailer is also a massive climate polluter: according to its own 2021 climate sustainability report, their carbon emissions increased by 18 percent over 2020. But in 2019, some 1,700 workers at Amazon signed a pledge demanding that Amazon start taking its climate impact seriously (https://ydatu.com/02-17). They had three demands:

> Amazon must demonstrate real climate leadership by committing to the following:
> 1. Zero emissions by 2030: Pilot electric vehicles first in communities most impacted by our pollution
> 2. Zero custom Amazon Web Services (AWS) contracts for fossil fuel companies to accelerate oil and gas extraction

3. Zero funding for climate denying lobbyists and politicians

But that's not all: by signing the pledge, the employees also committed to protest their employer's lack of action on climate change by participating in a global walkout on September 20 of that year. A *walkout* is exactly what it sounds like: workers walk out of their workplaces in protest.

But there's something else to note about a walkout, which is that it's a form of strike. A *strike* is simply another kind of collective action workers can take. In fact, collectively withholding your labor — collectively *stopping work* — is possibly the most powerful form of action at a worker's disposal. When used intelligently, it can draw public attention to an important issue, which in turn can apply pressure to your employer. But it exerts internal pressure, too: when a group of workers withhold their labor, that slows productivity, which impacts their employer's bottom line. That pressure can, in turn, be used to win concessions from an employer.

In recent years, we've seen tech workers realize, time and again, the power of collectively withholding their labor — of walking out of their workplaces. Of *striking*. It happened again in June 2020, when the *New York Times* decided to publish an opinion piece written by a sitting United States senator. In the piece, the author called upon the then-president to deploy the military against American protestors — protesters who, I should note, were outraged by the murder of George Floyd, a Black man, by Minneapolis police officers. Put another way, the author — a powerful elected official — advocated for widespread violence against American citizens.

I will not link to the piece. But I will note that in response to its publication, employees across the *New York Times* newsroom mobilized to stage several protests. And two days after the essay was published, over *four hundred* NYT employees — including many of the *Times*'s tech workers — virtually walked out of the office.

Just to be clear: these protests didn't *start* with those tech workers. They were planned and led by the *Times*'s editorial staff, many of them Black, who were marching because they feared for their safety. And I mean that quite literally: they worried that their employer's decision to publish the piece had made their colleagues, especially their Black colleagues, unsafe.

With that said, I want to note that in their open letter to management, the *Times*'s tech workers explained their rationale for participating in the protests, and for walking off the job:

We affirm the concerns our editorial colleagues in Newsroom and Opinion have already shared: that publishing the Op-Ed piece by Tom Cotton endangered our staff — especially those on the ground reporting and our Black colleagues broadly — as well as people participating in protests. As non-editorial employees of the company, we are typically circumspect about expressing dissatisfaction with decision-making that affects our published content, but we feel that this particular circumstance warrants our action. *We are saddened that the organization and tools we have endeavored to build and grow were used to publish this piece.* (https://ydatu.com/02-18)

Emphasis mine, because I want to underline the significance of that point. I mean, yes: above all else, the tech workers who built the company's publishing tools protested the essay because it endangered the lives of the reporters they worked alongside, especially their Black coworkers. The article advanced a racist, violent argument — one that literally made their colleagues unsafe.

But just as importantly, the tech workers were protesting because they recognized that their labor helped make the senator's article possible. They built and maintained the content management systems that published the article, which brought that article to a wider audience. Walking off the job — withholding their labor — meant that their work couldn't contribute to further harm. But just as importantly, the walkouts could also call public attention to the fact that the newspaper's editorial process had broken down, making their colleagues and coworkers unsafe.

Something similar happened at Facebook that very same month, when the then-president of the United States took to social media to threaten civilian protestors in Minneapolis with military violence. Twitter quickly added a publicly visible note to the president's tweets, saying they violated the company's rules around posts that glorified violence; Facebook, however, opted to do nothing. "Personally, I have a visceral negative reaction to this kind of divisive and inflammatory rhetoric," Mark Zuckerberg said on his Facebook page. "But I'm responsible for reacting not just in my personal capacity but as the leader of an institution committed to free expression."

In response, Facebook employees began openly questioning their company's inaction, using internal discussion forums to do so. Dozens of employees publicly posted open criticism of their employer's stance; one Facebook engineer resigned publicly, saying, "I cannot stand by Facebook's continued refusal to act on the president's bigoted messages aimed at radicalizing the American public" (https://ydatu.com/02-19). And when their company *continued* to refuse to act, approximately four hundred workers staged a virtual walkout.

A common thread runs from GitHub workers' letter to their CEO, to the tech workers' walkouts at the *New York Times* and Facebook. At GitHub, it was a collectively signed letter; at Facebook and at the *Times*, tech workers walked out of their offices in protest. But in each case, these workers were demanding that their labor be used more ethically by their employers. And they were able to make those demands because of their numbers: by acting collectively, they could protest powerfully *and* safely.

Because sometimes, safety is the very thing we're fighting for.

SOLIDARITY AT SCALE

It's safe to say that 2018 was an eventful year for Google, to say the least.

In the beginning of that year, reports surfaced that Google was attempting to bid on contracts for the United States Pentagon. Specifically, they were pursuing a project code-named "Project Maven," which aimed to provide artificial intelligence that could interpret video images — which could, in turn, be used to improve the targeting of drone strikes (https://ydatu. com/02-20, subscription required). Shortly thereafter, various internal leaks revealed the existence of "Project Dragonfly," a secretive project to create a censored, trackable version of Google's search tool for release in China (https://ydatu.com/02-21). When these leaks surfaced in the press, the projects were roundly condemned by human rights organizations across the globe; Google's own employees loudly protested them as well, both internally and publicly.

And then, in the final days of October 2018, the *New York Times* reported that Google had protected three executives accused of sexual misconduct — including Andy Rubin, the creator of Android, Google's mobile operating system. According to the reporting, rather than firing the executives, Google paid each of them millions of dollars upon their departure, even though the company was not legally required to do so. According to the *New York Times*:

> Google could have fired Mr. Rubin and paid him little to nothing on the way out. Instead, the company handed him a $90 million exit package, paid in installments of about $2 million a month for four years, said two people with knowledge of the terms. The last payment is scheduled for next month. (https://ydatu.com/02-22, subscription required)

On November 1, 2018, one week after the *Times* exposé, *twenty thousand* Google employees and contractors walked out of the company's offices around the world. They pushed back from their desks in protest, stopped working, and walked out of their offices. This is, to date, the largest protest action in the history of the tech industry. If you'd like to get a sense of the scale of the day, you can find photos from the official Google Walkout account on Twitter (https://ydatu.com/02-23, registration required), or in various media outlets (https://ydatu.com/02-24). But according to the walkout's organizers, nearly 75 percent of Google's offices saw protests: the largest were in New York City, with three thousand participants; and at Google's Mountain View headquarters, which had four thousand participants (https://ydatu.com/02-25).

These walkouts were, simply, a tremendously inspiring sight. Heck, on days I feel overwhelmed or depressed (or both!) about the state of our industry, I'll rewatch video footage from that day. Imagine walking out of your office, standing alongside thousands of your coworkers, and loudly demanding something better — *fighting* for something better — from your employer.

For me, the Google walkout is one of the most remarkable things I've ever seen in the tech industry. I mean, yes, the scale of it is absolutely inspiring — but that's not why I return to it so frequently. I think of it often because to my mind, it's one of the finest examples of *solidarity* to come out of our industry.

If I'm honest, solidarity is a bit of a wiggly term for me to define. It's easier for me to identify it than to define it, if that makes sense. But I think solidarity is, at its heart, an expression of shared values between two or more people — and just as crucially, how those values *connect* us. It involves recognizing that if there's some *thing* that injures you, well, that thing is also an injury to me. And we'd best fix that thing. Together.

You probably have some experience with this at work, even if it's on a project-by-project basis. If you've ever shared knowledge at your job — whether giving a presentation, writing a tutorial, or training a coworker in something you've learned — you've invested your time in your coworkers. And that act is, at least in part, an act of communal care: you're helping your coworkers because doing so increases the number of people at work who can depend on one another. Sharing that knowledge quite literally improves the entire community.

I realize that for some tech workers, this idea of solidarity can feel like a difficult concept to accept. And I get it, truly. For me, I wonder if some of that difficulty lies in the fact that the idea runs counter to the stories the tech industry tells itself about "meritocracy": about the brilliant self-starters who struck out alone, and who achieved dizzying successes simply and solely *because* they were so brilliant and talented and motivated, all without help from anyone else. We've bought into this idea of the talented individualist so thoroughly that our stories tend to brush away all the privileges they enjoyed, or the assistance they received: the family that supported them, the wealth they inherited, the educational institutions they had access to. Somehow, our fables focus solely on the child, rather than on the village that raised them.

And that, I think, is what makes the idea of solidarity so powerful. It is an open, explicit acknowledgement that there are connections and dependencies between us as workers. It's quite possible you already see ways in which your work relies on others: that your work as an engineer can't begin without the designers on your team; or that you, as a designer, rely on input from researchers or content strategists; and so on. And if there are people or systems that hold one of your coworkers back — or actively harm them — how could that *not* impact your work? How could it *not* impact you, as a person?

The American labor leader Eugene Debs once said, "While there is a lower class I am in it, while there is a criminal element I am of it; while there is a soul in prison, I am not free" (https://ydatu.com/02-26). To my mind, that's a beautiful expression of what solidarity can be. Our common humanity binds us together, and motivates us to fight for something better: for a workplace, and a world, that's more just, fair, and kind.

But as nice a line as that is, it's important to note that solidarity can't simply be stated. It has to be *shown*. It has to be *demonstrated*. And that's what I find so inspiring about the Google walkouts. Twenty thousand tech workers took to the street because they were protesting how their employer reportedly protected several powerful men accused of sexual misconduct. During the walkout, participants publicly shared stories of the sexual and racial harassment they'd faced at work for *years*, citing countless instances in which Google's leadership refused to protect victims, choosing instead to side with the harassers.

We have to center those survivors and their stories, full stop. As we do so, though, I'd invite you to notice how many tech workers showed up — and spoke up — in *support* of those who were directly victimized at work. They may not have been harmed themselves, but they could stand in solidarity with their fellow tech workers, and fight alongside them to make their workplace safer.

In fact, solidarity shines through the five demands that Google's workers presented to their leadership as they walked out of their offices. The organizers' letter is available in full online (https://ydatu.com/02-25), but their demands are worth reading by themselves:

1. An end to Forced Arbitration in cases of harassment and discrimination.
2. A commitment to end pay and opportunity inequity.
3. A publicly disclosed sexual harassment transparency report.
4. A clear, uniform, globally inclusive process for reporting sexual misconduct safely and anonymously.
5. A commitment to elevate the Chief Diversity Officer to answer directly to the CEO and make recommendations directly to the Board of Directors. And, to appoint an Employee Representative to the Board.

Each of these demands represented the workers' calls for a safe working environment: for an end to the sexual harassment, discrimination, and systemic racism they experienced in their workplace. They were demanding that Google address long-standing equity issues around compensation and promotions, and ensure that sexual harassment was stamped out across the organization. And by demanding the appointment of an employee representative to Google's board, the protestors wanted to guarantee that the company would *remain* accountable to workers' demands.

There's more context for these five demands available in an op-ed the organizers published at *The Cut* (https://ydatu.com/02-27). And as I read through the piece, I'm struck by a quietly recurring theme (emphasis mine):

All employees and contract workers across the company *deserve to be safe.* Sadly, the executive team has demonstrated through their lack of meaningful action that our safety is not a priority.

This [commitment to end pay and opportunity inequity] must be accompanied by transparent data on the gender,

> race and ethnicity compensation gap, across both level and
> years of industry experience, accessible to all Google and
> Alphabet employees and contractors.
>
> The improved process should also be accessible to all:
> full-time employees, temporary employees, vendors, and
> contractors alike.

Google has a frankly massive workforce. And like most modern tech companies of their size, it relies on legions of workers who aren't direct employees of the company: they're contractors, or temporary employees, hired to Google through third-party firms. But on a daily basis, each of those tech workers — whether they're paid a salary or hourly wages; whether they're a full-time employee or a contractor — still *work* together. They share office spaces, meeting schedules, and project deadlines. They are, all of them, tech workers. They perform tech work *together*.

That's what I find so admirable about the walkout's demands. At every turn, the organizers underlined that these demands didn't just apply to a specific class of tech workers at Google. They weren't fighting for safety for only the most privileged "class" of workers at Google. Instead, they were fighting for demands to make *all* of Google's workers safe — including the contract workers who had walked out alongside them. And these demands reflect the awareness that what affects one of Google's workers affects them all.

Solidarity in word and in action.

FIGHTING FOR THE TECH INDUSTRY WE WERE PROMISED

These workers walked off the job not just because they were furious about Google's reported decision to reward Andy Rubin for sexual misconduct; the *Times* exposé was simply the straw that broke the camel's back. In other words, while the *Times*'s reporting triggered the worldwide protest, discontent with the company had been growing among its workers for a long, long time. Here's Claire Stapleton, one of the organizers of the Google Walkout:

A few days before the Walkout [...] I sent out an email with a dumb-simple prompt: Why are you walking out? 350 responses came back. The Walkout's spark might have been Andy Rubin, and indeed there were plenty of other tales of harassment and coercion at Google. But it was broader, deeper than that; this was a monument to disillusionment, capturing all sorts of anecdotes and reflections on a culture of discrimination, gaslighting, retaliation, ethical breaches, punitive managers, bad HR. If I could boil all these responses down to a single question, it might be: when did you first notice the gap between what you believed Google to be — progressive, equitable, fair, good — and what you actually see and experience every day? (https://ydatu. com/02-28)

I'm struck by how Stapleton described the gap between her employer's stated values and how Google actually conducted its business — and how it treated its workers. It reminded me of this piece by Ann Haeyoung, another Google employee-turned-tech organizer:

Whether it was the uncomfortable advances of a senior product manager or white coworkers making racist jokes about non-white teammates, harmful behavior was part of the fabric of the workplace. As I became exposed to more of the business, I also became disillusioned by the disconnect between Google's onetime motto — don't be evil — and how it sought business with the military and big oil companies, not to mention its blatant disregard for the privacy and safety of the people who use its products. (https://ydatu.com/02-29)

This disconnect between motto and reality isn't unique to Google. In fact, a number of the workers I interviewed shared similar stories, especially those working in civic tech or media — in organizations they'd describe as "mission-driven." They each described a wide disparity between how their employers talk about their values publicly and how it actually feels to *work* at those companies.

I spoke with Sharon Warner, who helped organize Nava United, the union at the civic tech agency Nava PBC. She shared some of the motivations behind the union's formation, like fighting for better working conditions, and addressing questions around compensation policy. But she added that for her, unionizing was an opportunity to take the company's stated values and translate them into specific working conditions. "I genuinely think that the union makes Nava better," she told me. "Nava has smart, caring people, and it will only make Nava better to have that perspective heard. I really believe that."

I can't help but think of the people who walked out of Google's offices in the same light. They were workers who loved their company — or, more specifically, the *promise* of their company. These workers were moved to protest because they wanted to close the countless gaps between Google's corporate values and the harms involved in simply working *at* Google. These workers wanted to do good work; they wanted their jobs, and their employer, to be *better*.

As the walkout's organizers said in their editorial:

> A company is nothing without its workers. From the moment we start at Google we're told that we aren't just employees; we're owners. Every person who walked out today is an owner, and the owners say: Time's up. (https://ydatu.com/02-27)

THE TECH INDUSTRY THAT ACTIVISM BUILT

On a personal note, I think this activism is the single most inspiring thing happening in our industry right now. As individuals, these people are incredibly courageous, risking their livelihoods to fight for something better. But again, I want to point out that what we're seeing are many, many, many cases of tech workers realizing they have incredible amounts of power when they act together.

In fact, one volunteer-led organization called Collective Action in Tech has been documenting the modern tech industry's labor movement. They've published a remarkable public

---

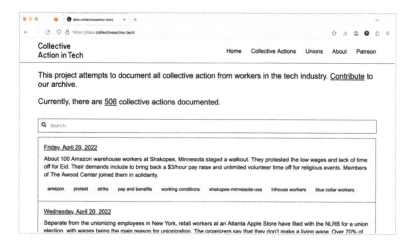

FIG 2.4: Collective Action in Tech's archive of — you guessed it — collective actions in the tech industry, is a truly expansive history (https://ydatu.com/02-31). Know of an action that's not in the database? You can add it!

repository of collective actions in the tech industry, ranging from the 1960s to the present (**FIG 2.4**). Thanks to their remarkable tagging scheme, you can quickly filter down their dataset to, say, find all tech industry protests that were led by white-collar tech workers (https://ydatu.com/02-30).

But if you spend a few minutes scrolling through the data, you'll quickly notice something: all of this good and inspiring activism we're seeing right now isn't *new* to the tech industry. Recent years have indeed seen a remarkable resurgence of tech worker activism — the protesting, the criticism, the mobilization — and at a scale the industry has never really experienced before. But it's equally important to note that tech workers have a deep, rich history of activism, a history that reaches back decades. And it's important to understand that *what's happening now* would not have been possible without *what happened back then*.

For one example, there's Computer People for Peace, an organization formed in the late 1960s. CPP is one of the earliest examples of tech worker activism I'm aware of — and I only learned about it in the last few years. Born out of the antiwar movement, the group created a national network of tech workers who opposed their industry's involvement in Vietnam, and in defense contracting more broadly. They argued:

> Computers are increasingly being used as a means of oppression. They are at the heart of every military and police system. They are at the core of every major corporation and are used to maximize profits with little regard for human needs. (https://ydatu.com/02-32, PDF)

CPP fought against what they termed "corporate racism" in the tech industry, protesting IBM's business dealings in apartheid-era South Africa; they provided testimony to the United States Congress, arguing for the need for data privacy at the federal level; they sued to stop the New York City Police Department's efforts at surveilling antiwar activists. CPP also intermittently published a newsletter called *Interrupt*, in which they advanced their vision for a better tech industry and announced new campaigns or protests (FIG 2.5). And they did all this within the span of seven or eight short years: from the late 1960s, until they disbanded in 1974.

As I said, I only learned of Computer People for Peace a few years ago, when the *Outline* published a short feature on the organization and its work (https://ydatu.com/02-34). And to be perfectly honest, I got a little upset. I felt like I'd stumbled upon something that'd been hidden from me: a piece of history that didn't square with the rhetoric I'd been told about my industry's lofty ideals. Here was an organization of tech workers who were fighting for racial justice, for gender equality, for workers' rights, for a demilitarized tech industry—and all of those fights were happening *decades* ago. It felt like CPP had been erased from history somehow.

Of course, that isn't what happened. There was no grand conspiracy to rewrite our industry's history books. Instead, it's more that our industry has always had a fraught relationship with the past. So much of our work is focused on what's *next*—the latest design tool, the newest technical framework, the next deadline—that we rarely stop to look at the *back then*.

1984 is here

The following call has been issued to peace and activist groups. In addition we urge all computer people to join us in Atlantic City in May.

Computers are increasingly being used as a means of oppression. They are at the heart of every military and police system. They are at the core of every major corporation and are used to maximize profits with little regard for human needs.

The Spring Joint Computer Conference (SJCC) is an annual trade show-technical conference-public relations gimmick-sales event which brings together representatives of major corporations (IBM, GE, Honeywell, RCA, Litton, Rand, AT&T, etc.), high level representatives of the military and government, and the technocratic elite that serves their interests.

Obviously the event is overwhelmingly dominated by white males.

SJCC is being held at Convention Hall in Atlantic City, N.J., on May 18-20. Attendance is expected to exceed 30,000, making the conference one of the largest military-industrial gatherings in the country.

of actions, meetings, and demonstrations during the SJCC. The issues to be raised include:

- US genocide in South East Asia, particularly corporate involvement. (Honeywell is the prime manufacturer of anti-personnel fragmentation bombs.)

- Repression at home, specifically the use of computer based information systems as a means of social control. (Military Intelligence keeps data banks on civilians--including all of us.)

- Corporate racism (IBM plans to expand its South African market while the rate of unemployment among Third World people in the US continues to increase.)

- The present misuse vs. the constructive potential of computer technology (as applied to health, education, welfare, housing, ecology, and urban planning).

FIG 2.5: An excerpt from the February 1971 issue of *Interrupt*, the newsletter by Computer People for Peace (https://ydatu.com/02-33).

That's one of the many reasons I'm grateful to the Tech Workers Coalition for their free newsletter (https://ydatu.com/02-35). In addition to publishing stories from tech workers across the industry, the volunteers behind the TWC Newsletter have been unearthing our industry's long history of activism and labor organizing (FIG 2.6).

Why do these stories matter? Well, without them, it's easy to look at the industry's past few years, and think that all of this activism is something new or novel; that workers fighting for a better industry are somehow *out of step* with how the industry has historically operated.

But when we look back to these stories, we can see that worker-led activism *was* how the tech industry historically operated. We can hear Dr. Richard Hudson and Marceline Donaldson,

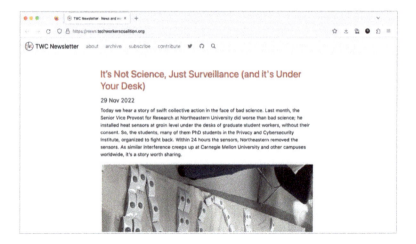

FIG 2.6: The Tech Workers Coalition's newsletter — readable online or delivered directly to your inbox (https://ydatu.com/02-35).

both of them Black employees at IBM in the 1970s, talk about the structural disenfranchisement they faced as workers at the tech giant (https://ydatu.com/02-36). We can read Karen Estevenin's account of her organizing work with the Washington Alliance of Technology Workers, or WashTech, which started as an attempt to unionize Microsoft in the late 1990s, before growing into an alliance of temporary and contract workers across Microsoft, Amazon, AT&T, and other tech companies — all of them fighting together for better pay and working conditions. As Estevenin said:

> Organizing was a little different before the rise of social media platforms. I learned how to build worker power with a group in my living room, with a pencil and legal pad. But many of the issues that workers face today, like concerns over employer-sponsored visas that make workers vulnerable and nervous about retaliation for organizing and lack of transparency around pay equity, were present in the dotcom days, too. (https://ydatu.com/02-37)

I'll be honest: reading these historical accounts gives me a bit of a chill. It's more than a little dispiriting to hear that these issues — the structural racism; the widespread pay disparity; the industry's reliance on hourly, underpaid contractors; corporations obsessed with chasing military contracts — are as old as the tech industry. But it's also instructive to know that *these issues are not new*. The scale of our industry's problems may have changed, but the shape's just the same. These issues have been here since the very beginning.

And so have the workers fighting for a better version of the tech industry. That's the other benefit we gain by reading their stories: we can learn from them, and better prepare ourselves for the fights ahead.

THE POWER WE BUILD

But even though activism from tech workers isn't new, there are a few things that are. Let's look again at the 2018 Google walkouts. The *Times* exposé was published at the end of November; one week later, twenty thousand employees walked off the job. It sounds remarkable — and it is. Truly.

But a walkout of that scale couldn't have been orchestrated in a single week. Or rather, it couldn't have happened without everything that'd happened over the prior year. As you might recall, Project Maven and Project Dragonfly, Google's early forays into defense contracting, were met with incredible internal resistance — led by Google's own workforce. In other words, Google employees had been mobilizing since 2017, fully *one year* before the 2018 walkouts.

And according to reporting done by Jacobin (https://ydatu.com/02-38), that mobilization took many forms, including the following:

- Groups of employees met in chat rooms to share their concerns with one another.
- One group penned an open letter and circulated it among employees for signature, asking Google's CEO to cancel the Project Maven contract outright.

- Another group got employees to submit questions to every single all-hands meeting, and then got *other* employees to upvote those questions to guarantee the questions would be asked during the public forum.

In other words, amid all this internal dissent, Google's employees built various kinds of *social infrastructure* — processes, workflows, and communication channels — that let the company's workers better connect and communicate with one another. They had been using that infrastructure to organize internally for over a year. And then, when the sexual harassment scandals came to light, Google's workers already had established lines of communication they could draw upon — as well as their justified anger — to organize a worldwide walkout.

And that verb is key. This is *organizing*. Organizing involves recognizing that, first and foremost, problems in the workplace require workers to band together, to collectively work to solve them. But to *solve* those problems? That requires a sustained, concerted effort. When Google's workers started speaking out against Project Maven and Project Dragonfly, they weren't planning to stage a global walkout. Rather, they were looking for ways to connect with one another, to protest more effectively, and to enlist their coworkers in various internal collective actions. But over time, and with a considerable amount of effort, those connections between coworkers, and those lines of communication, turned into something that could produce a global work stoppage.

FROM ACTIVISM TO ORGANIZING TO UNIONIZING

In this chapter, we've looked at the resurgence of collective actions in the tech industry. And I don't know about you, but from where *I* sit, the last few years have felt like quite a snowball effect. Workers have taken progressively grander collective actions, demanding safer working environments, demanding more accountability from their employers, and demanding a say in determining what their work should be used *for*. From the employee-penned open letters at GitHub and Amazon, to the worker-led walkouts at the *New York Times*, Facebook, and Google, workers have realized that, collectively, they have power: power to demand real change not just from their employers, but from the industry as a whole. Each time, those workers have learned their power comes from standing in solidarity with one another as tech workers — and from being *organized* as they do so.

But in the last few years, that organizing has moved into a new phase: tech workers have formed unions.

I say "phase" quite deliberately. The emergence of tech unions *is* new — but in a deeper sense, it isn't. Instead, I'd argue that these unions are the natural outcropping of the last few years of tech worker activism.

If we look at the mission statements for these unions, it doesn't take long to see how many of them are grounded in matters of ethics and fairness — it is, quite literally, the language of activism. In doing so, these unions — and the workers who formed them — are drawing a strong, bright line through decades of tech worker activism, reaching back to the fights waged by tech workers at the industry's very beginning.

Want an example? Let's look at the organizing principles for Kickstarter United, the union of Kickstarter's tech workers:

> On February 18th, 2020, Kickstarter employees voted to form a union, becoming the first major tech company in the United States to do so. We took this step in order to invest deeply in our collective future at the company we love, to create a workplace that is safer and more equitable for all, and to support organizing efforts across the tech industry and the many creative communities our work touches. We are proud to call ourselves members of OPEIU Local 153. (https://ydatu.com/02-39)

And next, there are the values behind the formation of Act-Blue's Tech Workers Union:

> We seek workplace practices that are transparent, people-centered, and responsive to the diverse needs of our vibrant community both present and future. We seek ethical standards in our products that preserve trust and protect the vulnerable. (https://ydatu.com/02-40)

On the face of it, these are strong statements of principle — of the *moral priorities* these unions bring to their organizing. At ActBlue, they're explicitly calling for "ethical standards in our products that preserve trust and protect the vulnerable"; at Kickstarter, they're fighting for "a workplace that is safer and more equitable for all." These workers are demanding a more ethical version — a *better* version — of their respective companies. And a union is how they're going to get it.

Now, what *is* a union, you ask? Well, answering that question is going to be fun.

So let's have some fun.

3 WHAT IS A UNION?

"There will be no prisons, no scaffolds, no children in factories, no girls driven on the street to earn their bread, in the day when there shall be "Bread for all, and Roses too."

—Helen M. Todd, "Get Out The Vote," *The American Magazine*, 1911 (https://ydatu. com/03-01)

THIS CHAPTER'S TITLE ASKS a big question, and I'll do my level best to answer it. But to do so, I need to start small. I'll tell you about a river.

(Okay, make that two rivers. But they're good ones, I promise.)

Over the past few years, as my world's gotten smaller and I've been traveling less, I've been trying to spend more time next to water. When I go out running, my favorite routes bring me alongside the Charles River. If you haven't seen it for your-self, you really should visit. Well, if you've been to the Missis-sippi, the Mekong, or the Nile — you know, a *proper* river — the Charles probably won't impress you all that much. But to my eye, the river has a beautiful, languorous shape. When I'm away from it, I miss it.

Its name wasn't originally "Charles," of course; English colonizers and their king had something to do with that. Before them, the indigenous Massachusett people called the river Quinobequin, which has been variously and descriptively interpreted as "bending river," "river that turns back on itself," or — my personal favorite — "meandering." I think each of those translations suits the Charles perfectly: if you're looking at a map, the river traces a long, curved line of slow, largely still water across the eastern part of Massachusetts.

But after its name changed, the Charles became anything *but* meandering. Its current was diverted to mills and factories, which drove Boston's early industrial explosion. And it *was* an explosion: the first British-style textile factory arrived in New England in 1793; a little more than twenty years later, 140 similar mills were believed to be operating within thirty miles of that first factory. By any metric, that is a fractally wild level of growth, especially in such a short span of time.

It's not like this happened by accident, either. This industrial revolution was driven, at least in part, by the arrival of the cotton gin in the United States, which supercharged the mass production of cotton on Southern slave plantations. And that cotton flowed north, feeding the growth of the textile industry and its many, many mills and factories. It's helpful for me to think about these two economies — cotton harvesting in the South, and its industrial processing in the North — working side by side, acting as twin engines for the United States' industrial development in the nineteenth century.

TOWARD A ROUGH DEFINITION

In both of these early industrial economies, American mass production relied on an *exploited labor force*, in which employers' profits came at the expense of their workers. In the North, near the rivers and forests I call home, the expansion of mill towns relied on low-paid workers to labor in the mills: many of the workers were women and children; many of them were immigrants. In the South, the cotton industry relied on the enslavement of Black people, literally selling humans into bondage to toil under dehumanizing, deadly conditions.

Put another way, the expansion of these twin industries was founded upon deep, structural injustices. But time and again, those injustices were repeatedly resisted by workers who banded together to fight for something fairer — something *better*.

Want a couple examples? Here you go:

- One constant for New England factory workers was the long, grueling days they were forced to work — anywhere from twelve to fifteen hours a day, with only a handful of holidays each year. And mill owners would fire any worker who refused to submit to those strenuous schedules. However, that didn't stop the workers from pushing back: in fact, New England mill workers are believed to have invented the practice of overtime pay in 1817, demanding extra wages if they were asked to work more than seventy-two hours in a week (https://ydatu.com/03-02).
- Three decades later, factory workers across New England led several loosely coordinated campaigns to demand that laws be passed to limit the American workday to ten hours. The campaigners circulated petitions throughout the region, collected thousands of signatures from workers — most of them from women — and gave testimony before the United States Congress on the direness of their working conditions (https://ydatu.com/03-03).

In a very real sense, each of these groups was a *union*. These workers identified something about their work that was unjust, something that they wished was different — or something they wanted to preserve — and they acted together to change it.

And we saw in Chapter 2, that "acted together" bit is key. These workers weren't operating as individuals; rather, they took *collective action*. Because when one worker asks for better pay, their impact is limited: they might receive their raise; they might be denied their request; or they might be dismissed outright. But when a group of workers demands that raise, that changes the math somewhat. It's much harder to fire twenty workers making a request. Harder still to fire two hundred — or two thousand.

But it's not just about the numbers: it's also about *organization*. A union identifies an issue, and then decides *how and when* to act on it. In 1866, Black washerwomen in Jackson, Mississippi, collectively wrote an open letter to their mayor, in which they protested the low wages they received from white families (https://ydatu.com/03-04). And in that letter, they also announced their intent to offer a uniform rate as a group, one that more fairly compensated them for their work.

> Be it resolved by the washerwomen of this city and county, That on and after the foregoing date, we join in charging a uniform rate for our labor [...] and any one belonging to the class of washerwomen, violating this, shall be liable to a fine regulated by the class. We do not wish in the least to charge exorbitant prices, but desire to be able to live comfortably if possible from the fruits of our labor. (https://ydatu.com/03-05, PDF)

The letter couldn't have happened if the washerwomen hadn't decided *as a group* that their pay was inadequate. Once they did, they agreed *as a group* upon a new rate: $1.50 for a day's washing; $10.00 per month for individuals; $15.00 per month for families. From there, they decided — again, *as a group* — on the best action to take, and resolved to send their letter. But that action happened once they established that consensus on how to act: once they had *organized*.

Thinking back to the question at the start of the chapter — *What is a union?* — let's review what we've got so far.

> A union is a group of workers who, through organization and collective action, fight for a better life.

That group of workers could be two people; it could be two hundred. But organization and action are what helps them move together toward something better.

Now, this is a rough working definition of a "union." But I think it's a good start. Let's keep going, and see how well it holds up.

THE WORKERS WHO FOUGHT
THE MILLS OF LAWRENCE

Ever since industrialization came to the United States's shores, workers have, time and again, taken action to try to make things better for themselves: to improve their pay, to make their working conditions safer, to make their work fairer and more equitable. And as they did so, they made things better for *us*. Workers fought for twelve-hour workdays, then fought to lower that to ten hours, then to eight; they fought first for six-day workweeks, then five. They fought for sick leave, for minimum wages, for the idea of weekends — and for many, many benefits we take for granted today. In very real, practical terms, the history of organized labor is the history of progress.

That's not to suggest that progress came easily. For every one of these workers, standing up to their bosses was a *fight*. (And honestly, it still is.) In fact, there's one powerful example of this near me, in Lawrence, Massachusetts. Lawrence is a small city a bit farther north from the Charles River, seated on the banks of the Merrimack River. And thanks to the Merrimack, Lawrence was home to a thriving textile industry at the start of the twentieth century. Nearly all of Lawrence's residents worked in its mills, most of them immigrant women and children. In fact, many of the mill workers were as young as fourteen — and some were even younger, lying about their age to be able to eat.

Life in these factory towns was, simply, bleak. Working conditions were deplorable, and death wasn't an uncommon end for a textile worker — nearly a third were said to die before they reached the age of twenty-five. In response, the Massachusetts state legislature passed a law in 1911 to lower the maximum workweek from fifty-six hours a week, down to fifty-four. The legislators felt that the two-hour reduction would relieve some of the burden from mill workers. The mill workers, on the other hand, worried that their employers would reduce their pay by two hours a week, and simply speed production to make up for lost time (https://ydatu.com/03-02).

And at least in Lawrence, those fears were borne out. The law took effect on January 1, 1912; when workers opened their paychecks on January 12, ten days later, they found they'd been docked two hours' wages. Almost immediately, they took to the streets. A number of Polish women were said to have been the first to turn off their looms, their cries of *"short pay! short pay!"* rising to the rafters as they walked out of the factories. And as they walked, more voices joined theirs. Eventually, some twenty thousand workers had filled the streets of Lawrence. They marched, they picketed, they sang — together.

All the while, they were *on strike*. As we saw in Chapter 2, a strike is a remarkably powerful form of direct action, possibly the most powerful one available to workers. And its power was on full display in Lawrence: by taking to the streets, the mill workers had effectively ground all production in the town's factories to a standstill.

Most of the workers were women, and many were immigrants — and nearly all of them lived and worked in abject poverty. But they didn't *just* want to be able to feed themselves; they wanted to be seen as worthy of respect, something all too often denied them as workers. One group of women was said to have carried a sign bearing words from the suffrage activist Helen Todd: "We want bread, and roses too." Their slogan was a call not just for basic subsistence, but for a life of dignity.

The mills' owners responded to that call brutally, and with force. The state militia was called into Lawrence to patrol the streets, and to break up meetings and marches; mass arrests followed shortly thereafter. In the middle of a New England winter, soldiers used hoses to spray protesters with freezing water. And in February, roughly six weeks after the strike began, a group of women went to Lawrence's railroad station, attempting to send their children to live with sympathetic families in other cities. Police surrounded the women and children, beat them viciously, and detained some thirty women in jail (https://ydatu.com/03-06).

Public sentiment had been souring on the mill owners' treatment of the strikers for some time, but this one scene was a flash point. Newspapers reported on the attack, which outraged the nation. Congressional hearings soon followed, and featured testimonies from workers and their children — all of which brought the mills' terrible working conditions into the public record.

Shortly afterward, the mill owners moved to settle the strike. On March 12, 1912, they agreed to the strikers' demands, giving the workers a 15 percent pay raise, guaranteed overtime pay, and amnesty for the strikers. Other mills in the Northeast quickly made similar offers to their workers, fearing similar strikes and scrutiny. And the mill workers of Lawrence returned to work, having decisively won their fight with strategy, with careful planning, and with community. Their victory saw them back at their looms — carrying not just their bread, but their roses, too.

BUILDING THE SUPPORT SYSTEMS

Most stories about Lawrence — about the Bread and Roses Strike — tend to focus on the first day of the walkout — including the story I just told you. Those stories might begin as mine did, by telling you about the brave women who left their looms, their cries of outrage, the people who joined them in the streets. It's a story of how these workers shut down an entire city's industry and, in doing so, how they *won*. It's a story of a fight that was desperately, brilliantly fought.

But it's not the whole story. The strikers didn't spontaneously walk out the moment they received their pay. They likely couldn't have: the workers of Lawrence's mills represented over fifty different nationalities, and dozens of different languages (https://ydatu.com/03-07). The act of simply *forming and communicating* a plan to one another would have required enormous amounts of time, effort, and translation.

Thankfully, Lawrence's workers *had* invested enormous amounts of time, effort, and translation. Years before they walked out of their factories, they had been working with organizers from the Industrial Workers of the World (IWW), a labor union founded in Chicago in 1905. What's more, many of Lawrence's immigrant workers had fought similar labor battles in their various countries of origin. And they put that experience to use in Lawrence, building cooperative spaces in which the mills' workers could gather, eat together, and share ideas.

In other words, the Bread and Roses Strike is a perfect example of the slow, steady pace of organizing. Just as with the walkouts at Google in 2018, the Lawrence workers' strike couldn't have happened without all the deliberate, careful outreach and planning they had already done. By the time the Massachusetts law was passed at the end of 1911, the mills' workers already expected their bosses would reduce their pay—and they had already been preparing for it. For years, the workers of Lawrence's mills had been building systems to care for each other—to feed each other, to *speak* to each other—as a community of workers. And when it came time for the workers to shut down Lawrence's mills, those systems were ready to support them through the strike ahead.

UNIONS, THE LAW, AND US

The reason I shared that story about the Bread and Roses Strike isn't just because I love the story. (Although I do. I really, really do.) I started with Lawrence because the mill owners' response is a fairly common example of how most union fights have been received throughout US history: with threats, intimidation, and outright violence. Business owners would call upon police or militias (or both) to attack striking workers; union organizers would be threatened, or run out of town—if they weren't killed outright. In other words, there were precious few legal protections for workers in this country, much less for workers who wanted to unionize. And employers exploited this lack of legal protection, fighting ferociously and viciously to preserve their own unilateral power in the workplace.

This changed during the Great Depression, a time of dramatic economic and political upheaval. Financial markets imploded, creating worldwide instability, mass unemployment, and painfully tight labor markets across the globe. In the United States alone, millions of workers across the country struggled to provide for their families, which led to a wave of strikes across the country. And during this period, unions—and the promise of security they offered—saw a massive spike in popularity.

In response to this, Congress passed the National Labor Relations Act, or NLRA, in 1935. The act defined a host of protections for workers, and offered them critical tools for building power in the workplace. Just as importantly, the NLRA also created the National Labor Relations Board, or NLRB, which acts as a regulatory body for labor-related matters in the United States. It has several administrative functions, including overseeing union elections, but it also acts as an enforcement body: one empowered to prosecute and investigate labor law violations.

These are just a few of the protections afforded by the act, but they're the ones most relevant to this little book. (I'll note that you can read the whole NLRA online at https://ydatu. com/03-08, if that's of interest.) But in short, the NLRA is the overarching federal framework for labor law in the United States. It provides a significant number of protections for workers, whether their workplace is unionized or not.

A note on unions and their legal contexts

I realize that this chapter's had a decidedly American bent to it so far, diving as it did into some labor histories from the country I call home. But regardless of the country *you* call home, I'd bet these stories feel broadly familiar to you. After all, every country has its own stories of workers struggling together against injustice: the strikes of Scotland's Radical War in 1820; the Black-led unions in South Africa that held anti-apartheid strikes in the late 1980s; the 250 million workers in India who led a general, nationwide strike in 2020; the rail, education, healthcare, and postal strikes during the United Kingdom's 2022 cost-of-living crisis; and many more.

The cultural contexts and details differ dramatically from country to country, as do the workers' demands. But there's a universal story here — of ordinary workers recognizing that together, they have *power*. And if they're properly organized, those workers can collectively use that power to fight for what they need, and what they want: a better life. Regardless of where you live, these themes are pretty universal.

That said, a union also exists in a *legal* context: specifically, a country's labor laws define the protections unions are legally allowed to offer workers, while also defining the ways in which unions can legally operate. That's why I'm going to keep this book's focus on the United States, the country in which I write. Because in practical, legal terms, a "union" *does* mean something different in every country. Again, that's not to say there aren't considerable points of overlap—in fact, there are more similarities than differences.

Oh look, the United States discovered labor protections

With the NLRA's passage, the path to forming a union was formalized in the American legal code for the very first time. More specifically, that path was *legally protected* for the very first time. Employers could no longer retaliate against their employees for attempting to form unions. Additionally, it was *illegal* for employers to refuse to negotiate with their employees' unions.

The National Labor Relations Act has been modified significantly since its passage in 1935, by both Congress and the courts. (We'll talk more about this later.) And it's worth noting that the NLRA was a *highly* controversial piece of legislation. Business owners and conservative politicians despised the NLRA, chafing against the checks placed on the unchecked power that employers had previously enjoyed. And some labor unions of the time saw the NLRA as a legislative compromise, arguing that the bill constrained the labor movement's strength by funneling union organizing into a management-friendly framework.

But even still, the NLRA is a landmark piece of labor legislation, and I recommend becoming familiar with it. It governs not just how unions are formed in the United States, but how they operate—and how they can protect you and your fellow workers. Understanding labor law and how it protects you (and how it doesn't) will make you a better, more effective organizer. In that spirit, let's take a brief look at two key sections of the act, and see how they work for you.

First, let's look at Section 7 of the NLRA, which is possibly the most important part of the bill. In fact, it's described by some as the cornerstone of American labor law. It begins:

Section 7. [§ 157.] Employees shall have the right to self-organization, to form, join, or assist labor organizations, to bargain collectively through representatives of their own choosing, and to engage in other concerted activities for the purpose of collective bargaining or other mutual aid or protection [...]

Arguably, this is the heart of the NLRA. This line enshrines some fundamental worker rights into US law for the very first time. More specifically, the NLRA defines a class of "concerted activities" — literally, a set of activities that workers take *in concert* with one another — to collectively improve the conditions of their employment. It's true that elsewhere in the act, the scope of those activities is somewhat curtailed, limiting them to matters related to "wages, hours, or other working conditions." But even with those constraints, those concerted activities could include participating in a strike with your coworkers, discussing wages openly to address pay disparities, circulating a petition demanding better benefits, or collectively refusing to work in unsafe conditions. The fact that the NLRA defines and protects these activities is significant, because it ensures that you and your coworkers can engage in them without fear of retaliation from your employer.

The other section of the law I'd like to note is Section 9. It begins a bit drily, but it's just as crucial:

Section. 9 [§ 159.] (a) [Exclusive representatives; employees' adjustment of grievances directly with employer] Representatives designated or selected for the purposes of collective bargaining by the majority of the employees in a unit appropriate for such purposes, shall be the exclusive representatives of all the employees in such unit for the purposes of collective bargaining in respect to rates of pay, wages, hours of employment, or other conditions of employment [...]

Here, the NLRA names the right of workers to form labor unions, and defines a process for doing so: by conducting elections in their places of employment. Additionally, the act names the right of workers to perform *collective bargaining*: to use their unions to negotiate the terms of their employment, well, collectively. As a whole. The results of this negotiation process are written down in a *collective bargaining agreement*, or CBA. It has an admittedly fancy name, but a CBA is really just a contract: a legally binding contract between an employer and the union's members, one that formalizes how both parties work together.

Protection for me, and for thee

It's important to note that while these two sections of the NLRA are deeply related to each other, they're still very much separate legal concepts. Thanks to Section 7, you have a set of fundamental rights in the workplace, regardless of whether you join a union. If you and your coworkers want to compare your salaries to ensure you're all fairly compensated, you don't need to unionize to do that. By organizing around those issues, you — individually, as a single worker; but also collectively, as a group of workers — are protected under federal law.

But Section 9 of the NLRA *builds* on those protections. By forming a union, you and your fellow workers also gain the ability to enter into *collective bargaining*. This is the negotiation process by which workers, operating through their union, sit down with their managers to define the terms of their employment. Those terms of employment could include wages, hours, vacation, health insurance, policies for promotion, benefits, and more.

The right to perform legally protected concerted activities, and the right to form unions: these are just two planks of the NLRA, but they're the ideas and protections most relevant to this book. Both of these concepts are built upon the idea of workers operating together — and more specifically, *operating together to protect each other*. After all, Section 7 says that you and your coworkers "shall have the right [...] to engage in other concerted activities for the purpose of collective bargaining or *other mutual aid or protection*." Emphasis mine, because this really underscores the extent to which labor law in this country is built upon the idea of collective care.

Take another look at the phrase I highlighted: "mutual aid and protection." You and I travel through many different sys-

FIG 3.1: If you're interested in learning more about mutual aid projects, this repository on It's Going Down has countless examples from North America, divided by region, state, and province (https://ydatu.com/03-09).

tems on a daily basis: various healthcare systems; our jobs; the billing and repayment systems for a student loan; your state or national governments; heck, even capitalism writ large. As a concept, *mutual aid* acknowledges that those systems don't adequately provide care for the people who live, work, and operate within them each day. In other words, those systems don't care for *us*. That's why mutual aid is founded on the idea that we—you and I, and everyone else who passes through those systems—can and should support each other, directly. In doing so, we can properly care for one another, and protect ourselves from the systems that don't.

Mutual aid isn't charity, which involves a one-way distribution of resources from a donor to a recipient, channeled through a charitable organization. Instead, mutual aid is *cooperation*, informed and led by the community itself. Over the past few years of the COVID-19 pandemic, we've seen countless examples of mutual aid in action (**FIG 3.1**). In the early days of 2020, there were stories of community groups collecting donations of face masks and sanitizers, and then distributing them to those who most needed them. Other communities began pooling resources to help purchase groceries, and then gave them freely to any neighbors who needed them. Mutual aid is about recognizing we're best able to care for ourselves by *collectively* caring for each other, and then taking steps to do just that.

That's why it's so significant that this idea of "mutual aid and protection" sits at the heart of American labor law: these concepts are, in a very real sense, what make your union so powerful. Your union *is* a form of mutual aid.

I want to sit with that point for a minute, and I'd like to invite you to join me.

A union's power isn't derived from its individual members, from its size, or even from its goals — or rather, its power isn't *only* derived from those things. By establishing a union, you and your fellow workers acknowledge that the well-being of every union member is bound up in the well-being of every single other union member. And that acknowledgment is the heart of a union in a very real, practical sense. A union is powerful precisely because of that interconnectedness between workers — that solidarity.

But remember what we discussed in Chapter 2: solidarity has to be demonstrated; it has to be expressed through *action*. For you, that action might mean talking to coworkers about the importance of a union, or by lending a supportive ear to a fellow union member about issues they're facing in the workplace; it could mean showing up for your union's weekly organizing meetings, or standing alongside your union during a walkout. (Or it could be *all* of those things.)

These actions are how you demonstrate your solidarity to your fellow workers, to the broader public — and just as importantly, to your bosses. Solidarity between workers is quite literally the engine that powers your union. And that engine is built collectively, over time, by those small but powerful public acts of care that you and your coworkers offer each other every day: by listening, talking, and showing up for each other.

A union is, at the end of the day, a system you're defining and building together as a community of workers — through organizing, conversation, fighting, and contract negotiations. By acting together *as a union* to achieve your shared goals, you can better protect and care for each other. It's precisely *because* you act to protect and care for each other that you gain power at work.

Let's take a look at how your union will help you do all that.

WHAT YOUR UNION DOES

We started this chapter with some historical examples of unions. And I realize that the story of work *then* might feel dramatically more severe, and much harsher, than how work feels *now*.

In fact, that was a common theme in the interviews I conducted with tech workers. Many people I spoke with admitted that when they first entered the industry, they regarded unions as an artifact of an earlier, rougher time. That's not to say they were necessarily *against* unions. But many of the folks I spoke with were brought up thinking of them as something that was important to older generations, working in more physically demanding fields — not as something relevant to the tech industry *today*, to people pecking away at computer keyboards in the twenty-first century. It wasn't something that folks felt was *relevant* to their work. Afton Cyrus, a member of America's Test Kitchen United, shared that in some of their earliest discussions about forming a union, she heard similar lines from her coworkers. "Some people were hesitant to think about it. Like, 'No, that seems extreme.' And 'Those are for the angry people.' Or 'Those are for, like, steel workers — that's not what we do.'"

But cast your mind back to these two questions I asked you, all the way back in Chapter 1:

1. For the things you like about your job: How would you keep them from changing?
2. For things you wish were different about your job: How would you change them?

As we discussed, no single person can provide answers to those questions. Instead, you're usually left to choose between different levels of uncertainty: either remaining at your current job, hoping it doesn't change for the worse; or leaving for a new job, which may or may not materially improve over your current one. Much like the mill workers of Lawrence showed us, there's very little you *as an individual worker* can do to make structural changes to your job. As one person, you're quite literally powerless to make change. This is precisely the problem that unions are designed to address. By forming a union, you and your coworkers can create a path to changing the terms of your employment: of preserving the things you and your fellow union members like about your job, and changing the things you believe should be different. In her book *A Collective Bargain,* the labor scholar and organizer Jane McAlevey puts it even more simply: "Unions are conduits for worker demands and fairness in the workplace." I really like this framing: that a union is a mechanism for you and your coworkers to get what you need. No more, no less.

Now, a union is able to provide that mechanism by putting three powerful tools at its members' disposal.

1. The first tool is *collective bargaining,* which we've already talked about a little. This is the process by which your union negotiates for a collective bargaining agreement, or CBA, which contains the agreed-upon terms of your employment.
2. The second tool is a *strike,* which is a process through which workers walk off the job to *withhold their labor.* By doing so, they can severely hamper production at their place of work, applying pressure on their employer to gain needed concessions.
3. The third tool is *democratic governance,* because a union is controlled directly by its membership — by the workers who belong to the union. This tool is, as you might imagine, less tactical than the other two tools, but it's just as important: it literally shapes every aspect of how your union operates. Workers vote to form a union in their workplace; once formed, members regularly vote in elections of union leadership; and a union's contract must be ratified and approved by — you guessed it — a vote by the union's membership.

Consider these three tools for a moment, and think about how *different* they are from how your job currently operates.

By offering a path to a contract, a union offers a fundamentally different vision for what a job can be. A union gives you options beyond simply leaving your job, or hoping it gets better; instead, it puts a mechanism in your hands that allows you and your fellow workers to help *shape* your job into something better. It's a mechanism through which you and your coworkers can collectively — and quite literally — build power.

I spoke to an organizer who worked at Mapbox during its employees' (sadly failed) campaign to form a union. And they shared something a coworker mentioned in one of their organizing meetings, about why Mapbox's workers deserved a union:

> We recognize the superiority of democracy in every part of life except at work. And it's interesting how poorly this is appreciated: that organizations make better decisions when more people make decisions, not when fewer people make decisions.

I've had this point playing on loop in my head ever since I heard it: *organizations make better decisions when more people make decisions*. What a wonderful concept. Wouldn't you like to be involved in making those decisions? Shouldn't you and your coworkers have a seat at that table? The NLRA provides you with important legal protections, it's true. But by forming a union, you and your coworkers gain more say over how you work: you can negotiate a collective bargaining agreement to formalize the power you all share and, in doing so, obtain more equal footing with your employer.

Of course, you might be wondering: What goes into a collective bargaining agreement, anyway?

WHAT'S IN A CONTRACT?

Forming your union is the first step; entering into collective bargaining is the second. The result of that process is a contract, formally known as a collective bargaining agreement (CBA). A CBA documents everything your union and your employer have agreed to during your negotiations. It could define pay levels for different jobs at your company, limit the use of monitoring or surveillance software at your job, or outline the health benefits that must be provided to all employees. It could define provisions to protect workers in case of layoffs, such as mandating the amount of severance pay and benefits you'll receive if you lose your job. In short, your contract should represent gains on the issues your union collectively cares most deeply about.

Your contract will also establish something called "the grievance and arbitration process," which defines how your union and management will resolve issues that come up during the course of the contract. These grievances could include disagreements between the parties on how to interpret sections of the CBA, such as when a worker feels that management has failed to live up to its side of the agreement. It outlines the process by which the worker, a representative from their union, and management will decide on a path forward, usually through a hearing—and how to appeal that decision if either side feels dissatisfied with the outcome.

Also, it's important to note that your collective bargaining agreement will have a life span. Literally. When the contract is finalized, it will have a date on which it expires. But before you reach that point, your union will sit down again with management to start negotiating a new CBA.

The employees of Kickstarter were among the first companies in the tech industry to form a union with collective bargaining rights, winning their union election in February 2020. After more than two years of negotiating with management, they finalized their first collective bargaining agreement. And they voted overwhelmingly to do so, with 97.6 percent of the entire Kickstarter United union voting to ratify the tentative agreement. Their contract truly is a milestone. And it's not just a milestone for them, but for our entire industry: it shows that unions *are* viable in the tech industry, and can result in contracts that protect our workers.

But then, the workers at Kickstarter United did something incredibly generous: they put the entire agreement online, making it available for other workers and unions to read (https://ydatu.com/03-10, PDF). And personally, I think Kickstarter United's CBA is a remarkably clear, accessible document; I'd encourage you to take a look at it, if you're interested. But even if you're not, the union outlined some top-line gains they're especially proud of:

- A guaranteed minimum 3% annual cost of living raise.
- Salary benchmarking based on a national average—no geographic pay bands.
- A profit-sharing bonus pool so that when Kickstarter thrives, we all benefit.
- An annual pay equity review to guard against any potential discriminatory pay practices.
- Guaranteed development frameworks for every role so that we know how to advance in our careers.
- "Just cause" provisions so that no one is disciplined or fired without clear, documented reason. (https://ydatu. com/03-11)

These are just the first six items, mind you; the full list is quite a bit longer. (And of course, the contract itself goes into considerably more detail.) But even this excerpt represents some remarkable wins by the union. They've established a regular cost-of-living increase, and removed discriminatory pay practices based on geography. They've also established a salary review process to make sure that fair pay at Kickstarter *stays* fair in the long term. And they've defined clearer frameworks for career development and advancement. Finally, they've won "just cause" provisions in their contract, ensuring that they can't be fired arbitrarily, at the random whim of their employers. In doing so, they've secured some stability for their union members. (And no small amount of peace of mind, I'd wager.)

Again, Kickstarter United's full CBA goes into considerably more detail than what I've excerpted. If you're curious about what a union contract looks like in the tech industry—and what a union contract could look like at *your* company—this document is a remarkable source of inspiration.

But even so, it's important to note that *your* union's first contract will likely look very different. I interviewed Jacky Alciné, a software engineer who has been involved in labor organizing for a few years now; he was an organizer at Glitch's union, and is currently a member of CfA Workers United, the union at Code for America (https://ydatu.com/03-12). We were talking about some brainstorming sessions he'd been in at both companies, in which he and his coworkers were dreaming up ideas for what they might include in a CBA—ideas people had thrown around for what their contracts *could* look like. Things like, "What if our parental leave policy lasted five years?" And: "Could our contract have mandated that 20 percent of every sprint was spent on tech debt?"

Alciné hastened to note that none of these proposals ever appeared at a bargaining table; they were just ideas that came up in brainstorming sessions. But that's what I think is so valuable about the collective bargaining process, and the discussions around them: that these discussions are meant to be *generative*. As tech workers, we're asked to work within a structure that prizes conformity and productivity: "At this company, this is how we work; therefore, this is how we expect *you* to work." We've seen how agitating for change at work is often expensive and risky—and occasionally dangerous. There's not a lot of room for imagination within tech work under capitalism.

THERE IS POWER IN A UNION

Unions offer an alternative. They provide us with levers, both legal and structural, that allow us to open up space for ourselves to imagine something different: more equitable salaries, a predictable level of stability in case you're laid off, more generous health benefits, a clearer path to promotion. What *would* your job feel like if you had more generous parental leave? What if your job provided you with a pension and retirement benefits? What if your company guaranteed gender-affirming care to its workers? Or full coverage of reproductive services? What would your company look like if a union member sat on its board? The process of organizing a union is an opportunity for you and your fellow workers to talk about not just what's wrong with work, but to imagine a form of work that is *better* for all of you. I can't think of anything more hopeful than that.

As we've discussed, unions aren't some bygone historical artifact from the early twentieth century, nor are unions only relevant to workers in other industries. Unions are for people who might be dealing with a boss who is unhelpful, if not actively hostile; and for people who work in conditions that are toxic, unsafe, or harmful. A union can provide protection from those harms — through both the collective bargaining agreement it provides, and through the people in your union who stand beside you when you're wronged.

But it's just as important to recognize that unions are for workers who *don't* fall into these categories. Maybe you have a good relationship with your manager; maybe your benefits are good; or maybe you like where you work. But what happens when that manager leaves, or when you get assigned to a new product team? What happens when your management mandates a change in your workplace, a change that makes you feel unsafe? In other words: how *temporary* are the benefits you enjoy at work? How quickly can they change?

Nozlee Samadzadeh, a member of the New York Times Tech Guild, put it beautifully when I spoke with her: "It's very common for unionizing efforts to start when something bad happens. But even when things are good, there's always something that happens to remind you that you don't have power in your workplace." Unions ensure that the things you like about your job can stay that way.

Even more broadly than that, unions provide you with a voice in your workplace — and they ensure you will *always* have a voice. The act of forming a union, bargaining for a contract, and standing together in solidarity: these are mechanisms for installing a democratic engine at your place of work. But just as crucially, they are mechanisms for connecting you to your fellow workers. That connection, just as much as the contract, is how your power is built and preserved.

Every worker deserves a union. That means you; that means me; that means all of us. Let's look at how you and your coworkers will win yours.

4 EVERY TECH WORKER DESERVES A UNION

"Unlike the broad-chested heroes of proletarian novels or Eisenstein films, rousing their workmates to rebellion with a single fiery speech, the classical rank-and-file organizer was more like a patient gardener [...]"

— Mike Davis, *Old Gods, New Enigmas* (https://ydatu.com/04-01)

I ALMOST DIDN'T GET a chance to sit down with Nora Keller, who works as a labor organizer. I say *almost didn't* because, well, my videoconferencing software decided to haul out some mysterious glitch just before our call. But I'm grateful we got past the bug, because Nora said something that stuck with me.

You see, before she became a full-time organizer, Nora was a product manager at the *New York Times*, and a member of the NYT's Tech Guild (https://ydatu.com/04-02). We'd been talking about how she got started with organizing, and how a lot of what she felt in those early days mirrors what most people feel when they get started. She said:

> The thing is, no one knows how to do this before they do it. I keep saying that to people. I'm like, "No one comes into this qualified." Most people have never organized a union.

Emphasis mine, because this was a thread that ran through most of my interviews. Most people I spoke with were first-time organizers, and a good number of them admitted to feeling some trepidation, anxiety, or uncertainty — primarily because of how new the experience was to them. "I haven't been in a union before, much less tried to start one." "I don't know how to start; what if I mess something up?"

But what I want to underscore is that these people *did* it. And that means you can, too.

In this chapter, we'll look at the process for forming a union. Well, "process" isn't the best word: this isn't a neat, orderly set of steps with a strict, sequential order. The path to a union is going to look a little different at every company, because your needs and your coworkers' needs are unique to where you work. Some of these steps can happen simultaneously, or in a different sequence altogether.

That's why I'd urge you to think of this chapter as a *rough sketch* of how your union will come into being. The beginnings and ends of this process are approximately the same at each company — but between those two points, there's a lot of variation. Much of what you'll read below will be relevant to where you work, and some of it you'll need to adapt. With that said, most of what I've laid out below should be relevant to you. And as we look at each step, we'll look at lessons that were shared with me by various tech workers and union staffers, and review some resources that might be useful to you.

Ultimately, I hope this chapter gives you a sense of what to expect as you and your coworkers build your union and that, in doing so, it makes you feel confident and assured about the work ahead of you. It's still a new thing, which can feel difficult or scary. But I hope that by seeing a rough sketch of the path ahead, you'll feel the road's a little easier to travel.

And besides, this is a *union*: by definition, you won't be traveling that path by yourself. None of this rests solely on your shoulders. You'll be supported by coworkers standing alongside you, as you all work together toward something better.

GETTING OUR BEARINGS

As we get ready to dive in, let's talk a little about what I'm covering here, and what I'm not.

There are two types of unions in the United States: *majority unions* and *pre-majority unions*. Broadly, both of them involve a group of workers banding together to fight for improvements in the workplace. But they each operate very differently: they have different powers under the law, and different relationships to the collective bargaining process.

Nearly every union you read about belongs to the first camp. Majority unions have access to the collective bargaining process and, by extension, the ability to negotiate contracts between the union and the company's management. If a simple majority of workers agree to form a union, then the union is formed. (Hence the name: *majority* union.) Once the majority union has been formed, every worker eligible to join the union, including those who voted "no," will be represented by it.

By contrast, pre-majority unions don't require an employer to recognize them. Instead, they simply require employees to decide to join the union. But under the NLRA, employers are *not* required to negotiate with pre-majority unions. In other words, members of these unions won't be able to collectively bargain for a contract.

There is some nuance here: in theory, a pre-majority union could insist that they have the right to collectively bargain with their employer if they have sufficient leverage — if the union's membership represents a sizable percentage of a company's workforce, say, or they work on critical systems. In that case, refusing to bargain could leave the company open to disruptions, especially if the union elects to strike. But generally, under current labor law, management is not obligated to sit down at the negotiating table with a pre-majority union.

For workers at some larger corporations, where it might be impossible to organize a massive workforce into a majority union, the pre-majority model offers a compelling alternative. But employees who oppose the union won't be represented by it. That's why pre-majority unions are sometimes called "members-only unions," "minority unions," or "solidarity unions":

these unions only represent the workers who elect to become union members.

This book is focused on the path for forming a *majority* union in your workplace. The collective bargaining process — and the legal weight of the contract it produces — is an incredible tool for workers both to protect themselves and to define what their job can do for *them*. That's the form of power I most want to talk about.

LET'S TAKE A QUICK LOOK AT A MAP

To get started, let's take a high-level look at how a union is formed.

A majority union represents a *bargaining unit*, which is a ten-dollar legal term for "the workers who are eligible to be represented by a union." And that's where you'll begin the process of forming your union: by identifying the workers in your company who can belong to your bargaining unit, speaking to each of them about the need for a union, and then gauging their support.

From there, the NLRA defines two possible paths to forming a majority union:

1. **A union election.** If at least 30 percent of the workers in a bargaining unit signal their support for a union, then the National Labor Relations Board will oversee a formal union election. And if a majority of the workers in a bargaining unit vote "yes," then the union will be formed.
2. **Voluntary recognition.** An employer may elect to voluntarily recognize a union, based on evidence of support provided by the union — say, by a majority of the unit's workers having signed union membership cards. If they interpret that evidence as an indication that their employees *want* a union, the employer may agree to recognize the union without the need for an election..

At that point, the collective bargaining process begins, with your company's management sitting down over many negotiating sessions with your union's bargaining committee. When both sides of the negotiating table agree to a draft version of a collective bargaining agreement, it has to be approved by a majority of your bargaining unit. Once it has been, your union will have its first contract in hand.

After your union has ratified the contract, workers in your unit will start paying *union dues*. These dues fund crucial union operations like strike funds, contract negotiations and enforcement, and communications. Some unions might structure their dues as a flat rate; others set their dues at a small percentage of your salary, often in the range of 1 to 3 percent. But however much you pay, and how frequently, these dues are literally how you and your coworkers sustain your union, ensuring it can continue to support you all.

Sound good? Great. Let's dive in, and look at the process in a bit more detail.

STEP 1: BUILDING SUPPORT

If you want to form a union at your company, there's one simple place to start: a conversation.

Remember our questions from Chapter 1:

1. What do you like about your job?
2. What would you change about your job, if you could?

The second question's often the catalyst for a lot of union campaigns. After all, chances are good that you're frustrated by several issues at your workplace. For many of the workers I spoke with, widespread pay disparity was a motivating factor: workers would discover they were paid significantly less than their peers due to discrepancies across departments — or frequently, due to their race or gender. (Or both.) For others, they wanted a clearer (and fairer) policy for promotions at work. Some workers told me they specifically wanted "just cause" protections at work, to ensure that they couldn't be fired capriciously.

Any number of these issues could be things you would want to change about your work — or maybe you'd change something altogether different. But whatever your issue is, here's the thing: I bet you're not the only one who feels this way. Start having a few quiet conversations with trusted coworkers. Try to sound out their opinion on the topic, and how strongly they feel about it. If they don't see it as an issue, that's good to know. But if they're just as passionate as you are that this is a problem, and that it needs to be fixed? Well, maybe ask them: "What if we formed a union?"

This is quite literally the starting point for your union campaign. Quiet conversations with friends and allies, imagining a version of your work that's fairer, more equitable, and *better*.

Forming your organizing committee

As those early conversations gain steam, you'll want to take the next step with your organizing. And if you've already had some early conversations with supportive friends and coworkers, you're probably well on your way to forming your union's *organizing committee*, or OC.

The OC's responsibility is to *organize the organizing*: to provide some structure and strategy as you build support for the union. Members of the committee will be talking to coworkers to get them to agree to join the union, and to sign union authorization cards; but they'll also be planning next steps, identifying the workers who need to be approached, and determining how best to plan and structure that outreach. The organizing committee tracks the current level of support for the union, and helps plan your union's strategy for eventually approaching management and taking the union campaign public. It's also responsible for ensuring that your supporters are motivated and engaged throughout the campaign.

Who's in your OC, you ask? Initially, it'll be the folks you speak with, well, *initially*. The first people who sign up for forming a union at your company are often the ones who commit to forming the OC. But more broadly, and as your organizing committee expands and evolves, you'll want it to include *highly motivated supporters* of your union: people who aren't just sold on the idea of a union, but who are ready to contribute their time and energy to make it happen.

Now, there's no standard size for an organizing committee. At the New York Times Tech Guild, they initially aimed for forty workers — that is, 10 percent of the four hundred workers they estimated were in their bargaining unit. That percentage was roughly consistent across the unions I interviewed, but it's ultimately up to you and your coworkers. Chelsea Noriega, one of the organizers at ActBlue Tech Workers Union, told me they decided to adopt a different approach: "We actually decided to go with a nontraditional structure, where everyone was sort of responsible for doing things." In other words, ActBlue's entire unit of tech workers was highly mobilized, so *they put everyone on the OC.*

Regardless of its size, your organizing committee will be responsible for strategy, planning, structure, and communications. You'll start setting up regular meetings — often weekly, but at whatever frequency feels right to you — to review the work you've done, and to plan the work that's left to do. In ways big and small, your organizing committee is the nerve center of your union campaign.

Talk to a union organizer

At some point, you should reach out to a union to help support your organizing campaign.

"Hang on," you quite sensibly ask, "why am I contacting a union if my coworkers and I are already forming a union of our own?"

You are asking some downright stellar questions today. (Also, if I may be so bold, that hat looks wonderful on you.) This step is, technically speaking, optional: you and your coworkers can absolutely organize and form a union by yourselves. And if you did so, you'd be considered an *independent* union.

But vanishingly few unions go this route, and I'll try to explain why.

First: from an administrative standpoint, there's a significant amount of overhead that goes into the simple act of running a union. Financial disclosures have to be made each year to the federal government, for one. There are also legal resources to consider, which most established unions offer in spades. During two separate union elections, the Retail, Wholesale and Department Store Union filed legal objections on behalf of unionizing Amazon factory workers in Bessemer,

Alabama, alleging that Amazon had illegally interfered with both union campaigns.

Second, winning an election as an unaffiliated union is exceedingly rare. One notable (and wildly inspiring) exception to this is the Amazon Labor Union (https://ydatu.com/04-03), which led a grassroots, worker-driven campaign to organize at Amazon's fulfillment center in New York City. They won, and they won *big*. But victories like this are unbelievably rare: historically, elections won by independent, unaffiliated unions have been the exception, not the rule.

That's why unions are almost always *affiliated* with a parent union. More than a few organizers described the relationship to me as a little like a nesting doll: once you win your election, your union will be affiliated with a union *local*, which is effectively...well, the local branch of a national union. This is where your union dues will go. And once you become dues-paying members, you'll be able to participate in your local's democratic governance, weighing in on decisions alongside other local members. What's more, your local is the conduit through which you'll access the administrative functions, staff, and resources the national union provides its members.

This is what I meant when I said it's *technically* optional to reach out to a union: you *can* go it alone, but it's exceedingly difficult to do so. That's why I'd recommend reaching out to a few unions as soon as you and your fellow organizers are ready, and talk to them about how they can help organize your workplace.

Now, as I write this book, there are two national unions that've been most active in the tech sector. Let's take a look at them in turn.

The first is the Communication Workers of America, or CWA (https://ydatu.com/04-04). CWA is one of the largest unions in the United States, representing workers across various industries such as telecommunications, banking, media, and — you guessed it — information technology. In early 2020, they launched their "Campaign to Organize Digital Employees," or CODE-CWA, allocating resources and organizers to the task of organizing workers across the technology and video game industries (FIG 4.1).

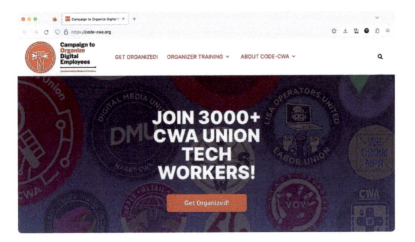

FIG 4.1: The homepage for CODE-CWA, CWA's initiative to unionize workers across the technology and video game industries, contains a simple message: "Join us."

CWA was involved in organizing Glitch's union, which was the first union in the tech industry to ratify a collective bargaining agreement (https://ydatu.com/04-05). Since then, they've only ramped up their efforts, partnering with tech workers at the New York Times Tech Guild, EveryAction, National Public Radio, Alphabet, and many more.

The other national union I'll mention is OPEIU, the Office and Professional Employees International Union (https://ydatu.com/04-06). OPEIU has a history of organizing and representing clerical workers, and worked on Kickstarter's landmark union campaign. And in 2020, they formed Tech Workers Union Local 1010 (https://ydatu.com/04-07), a new local specifically geared toward organizing in the tech industry (FIG 4.2).

I've had a few conversations with the Tech Local 1010 team, and I'm so inspired by their work. Not least because their win rate is *remarkable*: when I spoke to RV Dougherty from Tech Local 1010, they had won every single union campaign they helped organize. Dougherty attributed some of that to how Tech Local 1010 encourages sharing knowledge across their campaigns. There's a private online community for all of the

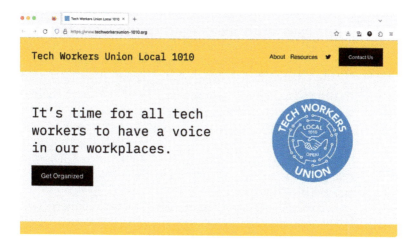

FIG 4.2: Tech Workers Union Local 1010 is the OPEIU-affiliated local that's helping to organize unions in the tech industry.

campaigns that Tech Local 1010 is involved with, in which organizers can ask each other questions, raise concerns or fears, and share strategies. In other words, it's a space in which unionizing workers can connect and support one another.

Now, there are many unions in the United States; CWA and OPEIU are just the two most active in the tech sector. And they're the unions most actively engaged in organizing tech workers *right now* — heck, by the time this book reaches you, there could very well be many more! If you find another national union that seems aligned with your union's values, and that understands the work you and your coworkers do, you should absolutely reach out to them. At this stage in your campaign, it's good to have options.

The process for selecting a union is ultimately up to you and your fellow organizers. Many workers I interviewed simply reached out to a union — or two, or three! — and struck up a conversation. From there, the unions can fill you in on what's ahead, and how they might be able to help.

That's not to say you can't make the search a union-wide project. Chelsea Noriega told me ActBlue's Tech Workers

Union approached it as a community research project, creating a #union-research channel in their Slack; from there, the membership collectively pulled information together on various unions. "We had people sort of going off and putting together proposals for different unions that had worked with tech workers before, and then we sort of evaluated them," Noriega told me. They evaluated hard data for each union ("How big is the union? How are their dues structured?"), but also interviewed workers they knew at companies who had already decided to affiliate with those unions. Equipped with that data, the organizing committee discussed their findings and made a decision.

Once you've selected a union, they'll provide your organizing campaign with a *union organizer*. These are people who have fought and won other union campaigns, and who know what to expect. Just to be clear, these organizers aren't running your campaign: they will offer guidance and advice, as well as training on how best to structure and organize your union campaign. Put another way, they're *supporting* your organizing campaign. They're going to work with your organizing committee to understand the issues you're fighting for, and to help you establish a plan for winning your union.

Mapping your organization (or, "Learning to love your spreadsheets")

We had been chatting about the digital tools she'd used on Mapbox's union campaign, when Bhavika Tekwani joked, "No one told me all this organizing would be just, like, *updating a spreadsheet*." But it's true: every organizer I spoke with mentioned that a spreadsheet was a vital tool in planning and managing every aspect of their union drive.

The process begins by identifying who you and your fellow organizers believe is eligible to be part of your bargaining unit — that is, creating a list of the workers you'd like to be in your union, along with their contact information. Here's what Afton Cyrus, from America's Test Kitchen United, told me about how their union started this process:

We created a clear list of who we thought the non-managers were across all departments, and we just put it in a Google spreadsheet, and then had fields for notes. Then we just incessantly tracked who had talked to who already, and what their level of support was.

To start making their spreadsheet, many organizers found it helpful to source a list of employees from online directories like those found in collaboration tools like Slack or Teams, or in corporate intranets. Once you've outlined your bargaining unit in a spreadsheet, that document acts as a map of your organization — more specifically, of all the workers in it. It should be treated like a living document: as workers leave your company, or new ones join, the spreadsheet should be updated to reflect that. The spreadsheet provides a picture of your bargaining unit's current size, and of its current level of support for your union.

But it's worth noting that this isn't a simple copy-and-paste job, because not everyone in your workplace can be part of your bargaining unit. What's more, those eligibility rules can differ dramatically from country to country. Here in the United States, the NLRA explicitly defines who can be considered a "worker" — and therefore, who is eligible to belong to your bargaining unit. If you work in the private sector in the US, you are considered to be a worker at a given company if:

- you are employed by the company,
- you are not an independent contractor, and
- you are not a manager or supervisor: that is, you have the ability to hire, fire, promote, or discipline other employees.

There are some additional exemptions you have to consider if you work in the public sector, or in specific industries (https:// ydatu.com/04-08). But practically, this all means that when you're trying to identify who can be part of your bargaining unit, you need to ensure that you're only including full-time employees who *aren't* supervisors.

(And of course, the tech industry seems to have an abundance of job titles that include the word "manager" — say, a project manager or a product manager—but do *not* have any supervisory responsibilities. So, hey. That's a fun little detail you'll have to plan around.)

Most organizers set up their spreadsheets in a format that can be edited collaboratively, usually in Google Sheets, along with information about each worker that's relevant to your union drive. Here's an example of what your spreadsheet might contain:

- The worker's name
- The worker's title, or their role at the company
- Nonwork contact information, usually a phone number and an email address
- Issues the worker cares about
- The union member responsible for talking with the worker
- The worker's level of interest in the union drive

This isn't an authoritative list, mind you: the list of fields you track can (and should) be customized to the needs of your campaign. Some organizers I interviewed limited their spreadsheet to contact information, level of support, and some free-form notes fields; a few others preferred more granular fields, which allowed for more complex searches and filters. But at minimum, your spreadsheet should keep a list of workers' contact information, as well as their level of support for the union.

And you'll find that the last item—their level of support for the union—will be especially important, because that single field will let you use the spreadsheet to track your organizing progress. Remember: you're looking to build support for your union across a majority of your fellow workers. As you and your fellow union members talk to more coworkers, you'll be asking them directly if they're interested in supporting a union at their workplace. The results of those conversations should get captured in your spreadsheet, since you're literally tracking how close you are to getting a majority of your bargaining unit to approve the union. This is effectively *the* metric of your campaign's success.

At most shops I spoke with, organizers used a 1 to 5 scale, which looked something like this:

1. Very pro-union
2. Pro-union
3. Undecided
4. Skeptical of unions
5. Anti-union
0. Uncategorized

(Yes, I realize that's a 0 to 5 scale. Yes, I took a philosophy course for my college math credit. No, I don't want to talk about it.)

The exact language used often varied across the unions I spoke with, but the general *shape* of the scale was basically the same. If a worker hadn't been spoken with, they were listed as "uncategorized" by default. But after someone reached out to that worker, the zero would change to some other number on the scale depending on how supportive — or unsupportive — they were of the union.

Over time, you'll find that the spreadsheet acts almost like a social map of your bargaining unit. Afton Cyrus told me:

> Every week on our [organizing] calls, we could pull up this spreadsheet and look at it together and say, "Okay, I could talk to this person; I know this person."

As you and your fellow union members conduct your outreach, it's helpful to draw upon any existing connections you have with individual workers in your company. Many organizers I spoke with found it helpful to start as Cyrus and the ATK United folks did: by talking to people they knew at the company, or with people on teams they were friendly with. Several people from the New York Times Tech Guild mentioned their spreadsheet also listed all the organizers who had previously reached out to a worker, just in case they wanted a little bit more background on how prior conversations had gone. If social connections can help you have more effective organizing conversations, you should absolutely capture that in your spreadsheet.

Keep it secure, to keep yourselves safe

I'm writing this book during a global pandemic, in which many tech companies have embraced hybrid or remote-first workplaces. That's been a boon for those of us who're working like hell to avoid wildly infectious viruses, but it *has* complicated some aspects of union organizing. After all, talking to coworkers about a union drive is best done *quietly* — and before we all worked remotely, grabbing coffee with a coworker was easier when you shared the same space. You could meet someone a block or two away from work, and quietly gauge their interest in a union, outside of earshot of your managers.

" — wait. Why do I have to talk about my union *quietly*," you pipe up to ask, "if organizing a union is protected by federal law?"

That is a marvelous question and, may I say, quite smartly asked! Sadly, the answer is both terrible and stupid: in the United States, labor law firmly favors employers. That might seem like a partisan statement, but it's a simple statement of fact.

What's more, that statement's been true for most of the history of organized labor in the US. Before the NLRA, business owners fought organizing workers with threats, lawsuits, arrests, and overt violence. And since the moment the NLRA was passed, business-aligned interests have been working to weaken or eliminate it. They've dedicated herculean amounts of time, money, and resources to whittling down the law's legal protections for workers. They first reaped a return on their investment with the passage of the Taft-Hartley Act in 1947, which dramatically undercut the scope and protections of the NLRA (https://ydatu.com/04-09). And in the intervening decades, the campaign to chip away at American labor law has only picked up steam.

On top of that, there is a massive and deep-pocketed industry built around helping companies bust unions, all while tiptoeing right up to the line of what's technically illegal — and in many cases, stepping over that line. Later in this chapter, we'll look at some ways in which union busting might manifest itself at your company. But in the meantime, it's best to assume that as soon as your bosses find out about your campaign, they will call in as much support as they need to stop it.

Again: much of this resistance is illegal. But in addition to the limits placed on the NLRA, the National Labor Relations Board—the agency tasked with enforcement of federal labor law—is perennially underfunded and understaffed. Of course, none of this happened by accident, as antilabor politicians have refused to provide the agency with the funding it needs to survive. According to a 2021 report, the agency lost roughly a quarter of its employees between 2010 and 2019 (https://ydatu.com/04-10), when adjusted for inflation, the NLRB's 2019 budget was actually *lower* than in it was in 2010. It's hard to have strong enforcement of labor law if your regulatory body is starved for the resources it needs to operate effectively.

So, yes. According to the law, it is technically illegal for your company to interfere with any organizing effort led by its workers. Legally, you cannot be fired for discussing your salary with your coworkers, or for discussing the need for a union at your workplace. If any of your NLRA-afforded rights are violated at work, you are entitled to file what's called an unfair labor practice charge against your employer, which is then investigated by the board. But due to the agency's lack of necessary funds and people, it can take months for an investigation to be concluded—if not longer.

This is why it's important to keep your campaign as quiet as possible, and to keep your company's management from finding out about it. For most organizers, that means getting all union-related communications off of company channels. Don't mention your union in Slack DMs, or over company email. Take your communications to phone, to private email addresses. Several organizers at ActBlue's Technical Workers Union told me that early organizing discussions all happened in one big group chat over Signal, the encrypted messaging app. But when the size of the union made that unwieldy, the union set up its own Slack instance, separate from the company-owned one.

STEP 2: OUTREACH AND ORGANIZING

At some point, your campaign will start talking to coworkers who *aren't* on your OC, and ask them to join the union. This is the time when you're working to build on the support you have, and to expand on it. You'll be approaching folks you work with — some of whom you know, but also many you don't — and trying to turn them into "yes" votes for your union. And you'll do that by having *organizing conversations*.

These might feel a little different from regular conversations, and your union organizer will help train you to effectively talk to your coworkers. But as an entry point, I quite like how the union organizer and labor scholar Jane McAlevey thinks about the relationship between *mobilizing* and *organizing*:

> Mobilizing is essentially doing a very good job at getting people off the couch who largely already agree with you. [...] Organizing, which I put the highest value on, is the process by which people come to change their opinions and change their views. Organizing is what I call "base expansion," meaning it's expanding either the political or the societal basis from which you can then later mobilize. What makes organizing different than all other kinds of activism is it puts you in direct contact every day with people who have no shared political values whatsoever. (https://ydatu.com/04-11)

I like how McAlevey sees these two actions as distinct, but deeply linked. In your organizing conversations, you're working to organize people to rally around an issue, which will allow you to mobilize them to *act* on that issue. The people you're speaking with may have different reasons for supporting a union — in fact, they might not agree with you politically, or at all. But in these organizing conversations, you'll be listening for areas of overlap, and then having an open conversation with them about how the union might help.

In the context of your union campaign, it might be useful to view these organizing conversations as having three parts:

1. First, you *listen* to your coworker's concerns, and work to understand them.
2. Second, you *link* those concerns back to your union's goals, and talk with them about how the union might help.
3. Finally, you make a *request* of your coworker. (In your early organizing conversations, you might be asking them to come to a union meeting; in the later stages of your campaign, you might ask them to sign a union authorization card.)

Shannon Turner, an organizer at ActionKit, said it's helpful to focus on the company in your organizing conversations: how it created an issue you're both struggling with, and most importantly, how a union can *resolve* that issue. She told me:

> It really does go back to active listening. Like, "What are your concerns? What are some of the frustrations that you have in the workplace?" Later on, you can connect what you hear to the union. Like, "I heard you talking about how frustrating it was that the support team is understaffed. With a union, we can actually fight for hiring more people, and for making sure that our workloads are actually sustainable. This is part of the reason why we're all coming together to form a union. Because right now, we know that this isn't working for us, and we want to make a change."

Of course, some people simply might not be willing to support the union campaign. If that's the case, just note their level of support in your spreadsheet, and keep working on your other outreach tasks.

I spoke with Brendan Zarkower, a technical product manager at the *New York Times*, and a member of the NYT Tech Guild. Brendan added one additional point that's critical to remember: "Your union represents the people that *didn't* vote for you, too. And it's very important to get their input." Just to underline that point: your union represents everyone in your bargaining unit, *including* the people who voted "no." That's another reason it's critical to connect with workers who might seem disinterested or skeptical — to look for areas of agreement, and work to understand why they disagree. It's import-

Joining together gives us a voice in making our workplace better for everyone. By completing the form below and clicking "submit," I am authorizing **Office and Professional Employees International Union** to represent me for the purpose of collective bargaining with my employer, **Bandcamp LLC.** My right to submit this authorization is protected by Federal law. I understand that this authorization card will not be shown to my employer.

Date _____

Print Name _____   Cell / Home Phone _____

Address _____

City _____ St. _____ Zip _____ Email _____

Work Email _____ Job Title _____

Signature _____

FIG 4.3: Your union authorization card might not look exactly like this sample card from OPEIU — it might even be digital! — but functionally, it'll be the same. Image courtesy of OPEIU (https://ydatu.com/04-06).

ant to continue to connect with them and mobilize them — not just during your organizing campaign, but *after* you win your union. (Besides, just because someone is disinclined to support the union today doesn't mean that that won't change in the future — so try to see that conversation as the first of many.)

As your campaign prepares to go public, you'll probably start talking to supporters about signing a *union authorization card*, which is...well, it's literally in the name. It's a short, modest-looking form printed on a card that asks for a signature and some contact information (FIG 4.3). The card basically serves two purposes: it not only signals the worker's membership in the union, but also empowers the union to collectively bargain for each worker who signs it. By signing the card, each worker *authorizes the union* to negotiate on their behalf.

Build support for now, and for later

As you're building support, it's easy to focus on numbers — counting rows in your spreadsheet, adding up the "yes" votes, and making sure your support is comfortably over the threshold you need to go public. And that's absolutely critical. But throughout this process, it's equally important to focus on your unit members as *people*, and ensuring they're as engaged as possible with the union.

There's a slightly mercenary-sounding reason behind this: namely, *is support for your union currently as strong as you think it is?* Someone might have been a solidly pro-union supporter when they were first added to your spreadsheet — but is that still the case? It's possible that circumstances at work have changed, or that they've simply changed their mind. In other words, it's helpful to think about the organizing campaign as not just gaining someone's support, but *keeping* it.

With that said, how you engage your members is up to you. Thankfully, you've got a lot of options. Regular union meetings can help. Several people I interviewed told me that their union Slack quickly became their favorite online space, full stop. Jax Engels, from the ActBlue Tech Workers Union, positively glowed when they told me about the community that the union provided: "It's so wonderful to feel like I have a group of people I can trust to fight for the things that I care about." And Kathy Zhang, a member of the NYT Tech Guild, told me that her time on the union is work — but it's not *all* work: "One of the nice things about organizing is that I've made a lot of really good friends through the effort, so it doesn't *feel* like extra work. It is kind of like a whole separate job, in a way, and there's parts of it that can feel really frustrating. But also, it can be really fun." That's another benefit to having a union-only Slack or Signal chat: it's a place for your union's community to flourish outside of work.

Additionally, your union can organize specific actions to help center your union's goals among the membership. For example, one of the NYT Tech Guild's driving issues was addressing pay inequality, since many tech workers were dramatically underpaid. They created a spreadsheet for workers to anonymously share their salary data — a federally-protected activity—and invited the unit to contribute. Over time, the spreadsheet helped strengthen the union's bonds:

> There was real hurt when people found out—as many did—that they were being underpaid. But the blame fell rightly to the employer, not fellow colleagues. And people in the channel were glad to hear about the highest salaries in each job level. Sharing them was a clear way for more privileged workers to support their marginalized colleagues, because the highest salaries in each job level showed the rest of us what it was possible to ask for. (https://ydatu. com/04-12)

Emphasis mine, because that point's key: the Tech Guild found that this activity helped remind people that a system underlies these inequalities—and that the union was going to address it.

But regardless of *how* you engage your members, it's important *that* you engage them. Anna Thorson, a former member of Catalist Union, told me what I'd heard time and again from organizers: that it's not just about numbers in a spreadsheet.

> It's not just enough to sign the cards. You want to show management of your company that it's not just names on paper: it's a group of people that are actually really, really invested in this thing they made.

Preserving your unit members' investment in the union is something that's worth...well, investing in.

Take care of yourself

Many of the organizers I spoke with were pretty honest about how much of their *everything* they put into their organizing. After all, organizing a union takes a considerable amount of work. It demands a lot of your time and energy, and there's no guarantee of success. It'll ask a lot of you on an emotional level, and you'll occasionally feel stressed, anxious, or both. Some of the folks I spoke with felt pretty burned out at different times throughout the process, and occasionally had to take breaks—or they had to step back altogether.

I asked nearly every organizer I spoke with how they balanced the demands of their full-time job with their *other* job of organizing a union. Here are a few tips I heard.

Trust your fellow union members

Nora Keller, the CWA organizer we met at the start of this chapter, told me repeatedly that asking for help is hard for most people—but it's incredibly important to do:

> It's important to delegate. Often, I think really motivated organizers try to do everything themselves. And it's really important to just push yourself to ask people to like, join the organizing committee, and to help them learn how to do stuff.

In other words, it's important to build resilience into the organizing you're doing. If you single-handedly own several key projects, those projects will be on pause if you need to step back. But if everyone in your union shares the load, the work can continue if anyone needs a break. ("Also," Nora added, "it's just not good for you! You're not going to be doing your best work if you're trying to do everything.")

It's okay to have limits

Whenever possible, make sure you're clearly communicating that breaks aren't just okay—they're *encouraged*. Vicki Crosson, a member of the NYT Tech Guild, told me she found that step incredibly important:

> We say up front [to new members], "You can stop at any time. You can leave for a few weeks, at any time. Just, you know, let us know what you need." And as a group, we try to make sure that somebody's checking in with that person after a couple of weeks. "How's everything going?" No pressure to rejoin, just, like, "how are you as a person?" And I think that explicitly making that space has been really helpful.

Emphasis mine, because I think that's an important point to underline. After all, you are one of your union's most valuable resources. If you feel like you need something from your union — a break, or more support on a project, or something else — make sure you advocate for it.

Celebrate your victories, no matter how small

The path to forming a union involves some big milestones, like winning your election and ratifying your first contract. But the path to those milestones is slow and arduous. That's why it's important to make sure you're celebrating each other, and celebrating the *little* milestones, too. If you manage to talk to everyone on your list in a given week, or get someone to agree to support the union? That's *huge*. And your union should celebrate that.

For example, Shannon Turner told me how ActionKit's union members celebrated as many wins as they could each week. "We made it a jubilant kind of atmosphere, where we would celebrate everyone. 'Yeah! You talked to all your people this week! That's amazing!' And really just trying to make sure that we uplift one another, and try to stay connected to why we're doing this."

Nora Keller, the CWA organizer who used to be part of the NYT Tech Guild, said something similar: "At the end of our organizing committee meetings, we would give kudos to people. We'd say, like, 'hey, like I want to affirm this person,' and point out something they did that was really great." Nora added, "Sometimes it's just a tiny win. Sometimes it's just, hey, I finally got this person to talk to me! Or sometimes it *is* a big win. But there's gonna be a lot of losses along the way. So be sure to celebrate the wins."

STEP 3: WINNING YOUR UNION

Once you've built enough support for your union, it's time for the next step. You're taking your union campaign from its underground status, and *going public*: you're about to approach management and ask them to voluntarily recognize your union.

When I asked various union members about how it felt to go public, every single person talked about how joyful the day was. There was plenty of organizing, to be clear: from writing the letter to management asking for recognition, to planning an all-unit meeting to make sure everyone knew the plan. But hell, it's a major milestone for your union—why *not* make it a party? ActBlue Tech Workers Union created a Spotify playlist filled with old union songs; and before they hopped on a call with management, they had an internal call to get people hyped for the meeting, and to celebrate what they'd achieved.

But some shared how nervous they felt, too. Anna Thorson told me that when Catalist's union of tech workers was about to ask for voluntary recognition, she felt scared. The union held an internal meeting before the call with management, and Anna said they talked as a group about what they were about to do. "We're gonna change our video backgrounds to show the union logo. And then we're going to jump into the meeting all together. And, you know, I gave a little speech about like, 'Let's be brave for each other.'"

As it happened, Anna ended up being one of the first to join the management call, and she shared how nerve-racking that felt. She told me: "I'm like, 'Man, if I'm out here alone, and nobody else is doing this, like, that's so bad!'" Honestly, I get it: How could you be on the verge of a big meeting with management, after months of hard, grueling organizing, and *not* feel nervous?

But once Anna told me what happened next, I got a little emotional:

> But then I started to see everyone blink onto the call, one after the other. And everybody had their union background up. It was just...it was the most powerful thing I've ever experienced. It was so many people showing that they had each other's backs. They were being so brave. These people didn't all have the connections or the career longevity that I had—this could be potentially a dangerous thing for them. But everyone showed up. It was really incredible.

When you go public, ideally your company's leadership will agree to *voluntarily recognize* your union. And for some companies, voluntarily recognizing their union is a business imperative. Here's how the CEO of Catalist announced the company's decision to recognize the union:

> Good communication and working relationships between staff and management are key to building a stronger organization to serve the progressive community. (https://ydatu. com/04-13)

Glitch, the first modern tech company to form a union, said that respect for their workers was critical to the company — and that supporting their right to unionize was important to building a better tech industry, too. Here's Zainab Shah, partnerships manager at Glitch:

> We are excited to set the standard for the tech community, and for us at Glitch to join CWA in pushing for ethical, humane, and responsible behavior from tech employers. We are also very excited to support workers everywhere engaged in similar struggles. (https://ydatu.com/04-14)

Assuming your leadership agrees to voluntarily recognize your union, congratulations! That means your union and management can sit down and start bargaining for your first contract.

But voluntary recognition is the *ideal* outcome — not the one you should prepare for. Your company's leadership may refuse to recognize your union, even after you approach them with a signed majority of union cards. Maybe they'll argue that the organizing you've been doing is somehow flawed, or that it isn't "democratic." Or that it's simply "not how we do things here."

Regardless of *what* they say, I need to be clear: you have a fight on your hands.

The management fight

I don't use the word *fight* lightly. In Chapter 3, we saw how the Lawrence mill workers' strike was met with a vicious, violent response from their employers. By and large — and there are still tragic exceptions to this — the modern response to union campaigns has gotten less physically violent. Instead, the response has become *economically and psychologically* violent. Workers have been disciplined for trying to organize a union, or have had their careers threatened; in some cases, they've been fired outright. Your company's leadership may wage a long, expensive campaign to delay your union certification, or to discredit it entirely. In other words, the modern management response to unions hasn't become more reasonable, or less severe. It just wears a slightly nicer suit, and doesn't call the cops.

The *kind* of fight you're facing depends on management's response. If they flatly refuse to engage with your union, then your union needs to start pressuring them to grant recognition, or to hold an election. If they agree to an election and set a date, then you've got a different set of problems: namely, they might be gearing up to wage an old-fashioned union-busting campaign, and to try to break your union's support.

With that in mind, let's look at some of the things your leadership might try during a management fight, and talk about how you and your fellow workers can get ready for them. I sincerely hope you won't need any part of this section, and can skip right over it. But as one worker told me, once they started seeing some of the tactics they'd prepared for, "it almost took the fear away, because we knew what to expect from them." As with any other part of this process, knowing what's ahead can help you be better prepared.

I'll start here: the playbook of tactics used during a union-busting campaign is incredibly old. If you look back at management fights over the twentieth century, you'll quickly find examples of these ancient, dusty tactics. But that doesn't mean you shouldn't take them seriously. Quite the opposite. The playbook hasn't changed much precisely *because* it's proven its ability to take the fight out of a union campaign, if not derail it altogether.

That's one of the reasons most union organizers will push your organizing committee to build *supermajority* support for your union: that is, getting support from a forcefully high percentage of "yes" votes — think 70 or 80 percent — rather than aiming for a *simple* majority of more than half of your bargaining unit. There's a fairly straightforward reason for setting the threshold so high: namely, union-busting tactics are often effective at eroding support, and at converting some of your "yes" votes to "no" votes.

But *"often* effective" doesn't mean *"always* effective." Your union members can prepare themselves by understanding these tactics and *why* they work — and then talk with one another openly about how best to counter them. In fact, your union organizer will have seen all of this before, so they'll have some strategies to help keep you prepared.

Hiring a union-busting firm

This is one of the early warning signs that you're going to have a fight on your hands. If your company's leadership is gearing up to try to bust your union, they will hire an expensive "union-avoidance" firm. These law firms will advise your company's leadership on how to quell any labor organizing at your company, and how best to derail your union campaign as quickly as possible.

Anti-union consultancies are an astonishingly big business — Amazon alone spent over four million dollars on these firms in 2021 (https://ydatu.com/04-15). And if you think *your* employer wouldn't go this route, it's important to realize that even ostensibly "progressive" companies will hire them: Google, Kickstarter, Starbucks, and Apple have each hired wildly expensive law firms to help them derail union campaigns at their companies (https://ydatu.com/04-16, https://ydatu.com/04-17, https://ydatu.com/04-18).

Again: it's technically illegal for your employer to interfere with *any* organizing effort at your workplace. But anti-union consultancies are hired because they have considerable experience operating just inside the line of what's permissible under American labor law. If your bosses decide to oppose your union, they'll be listening to these union busters very, very closely.

Captive audience meetings

If a union-busting firm ends up on your company's payroll, then *captive audience meetings* are likely going to start showing up on your calendars. They might not be called that specifically; they might instead be labeled "informational" or "educational" meetings. But whatever the title of the calendar invite, these meetings have one purpose: to try to pressure you and your coworkers to vote against forming a union.

In these sessions, you might be subjected to canned presentations or videos. Management might adopt a sympathetic tack: they might acknowledge some of the issues you're organizing around, and make promises about fixing them. (Or they'll make *noise* about fixing them in the future. Maybe.) Or you'll just hear some old-fashioned union propaganda, including scare lines like "unions are corrupt!" and "unions are after your paycheck!"

I spoke with Afton Cyrus from America's Test Kitchen United, who shared her experience of sitting through anti-union meetings. She told me the audience for these sessions *isn't* the highly motivated members of your union: these captive audience sessions are aimed at the members of your unit who might be less engaged, or at coworkers who might be a little uncertain. That's why these meetings exist: your management is trying to dissuade *them*, and to drive down support for the union.

Cyrus told me that's why it's so important to speak up in those meetings, as uncomfortable as it might feel. She told me, "Your job is to poke holes in their argument. You have to demonstrate that there's another side to what they're saying. And that means you have to get comfortable with being uncomfortable." For the workers of America's Test Kitchen United, they found it helpful to conduct a *mock* captive audience meeting for their entire bargaining unit. Cyrus told me the idea came from their union organizer, and added: "It was so key because [our union organizer] essentially came to the meeting as a quote-unquote 'manager,' and then taught us how to politely interrupt." As always, a little advance preparation can go a long way.

"Third-partying" your union

Your managers may start talking about your union as though it's an external entity. The language they use could be subtle, referring exclusively to "CWA" or "OPEIU," rather than the name of *your* union. Or they could adopt more dramatic rhetoric, talking about "union bosses coming in" to dictate your working conditions, or to change your benefits. (They'll probably mention union dues a few times here, for good measure.)

Additionally, you might hear them talk about "*the* union": that it'll be impossible for employees and managers to talk directly with one another, because "the union" will somehow get in the way. Or conversely, managers might refuse to talk to workers directly, saying that they will only negotiate with "the union" — that is, your union organizer, who doesn't even work at your company. (Look, I didn't say this line of thinking was logical.)

In fact, suggesting the union is an external party is, like much of the rest of the anti-union playbook, flatly untrue. Your union will be *affiliated* with a national union, but your union is *your union*. Your union was built by you and your coworkers. *Your union* organized around workplace issues that you and your coworkers are fighting to fix. Talking about the union "coming in from outside" is a tactic used to sow division, and to stir up questions and uncertainty among your bargaining unit. The direction of your union is democratically defined by its members — that's you and your coworkers. And nobody else.

So whenever you hear talk about "the union"? Make sure to quickly and loudly counter it with a correction: "it's *our* union."

Reducing the size of your bargaining unit

It's also possible that your management will try to pick a fight over who gets to *be* in your union. Specifically, they'll try to manipulate the size and shape of your bargaining unit. This could manifest itself in a number of ways:

- Management might formally contest some of the employees who've been marked as belonging to the unit, arguing that they're not eligible to be part of your union.
- Conversely, management might try to *increase* the size of your unit, either by adding workers from another department or by going on a hiring spree. Either way, they'll try to dilute your unit with (likely) anti-union workers, potentially spoiling your vote.

- Management might even promote members of your union into managerial roles, thereby making them ineligible to be part of the bargaining unit.

Some of these tactics can be fought with quick organizing and outreach. *All* of them can be fought with strong communication and tight bonds among your union members.

Creating fear, uncertainty, and doubt

Management's goal during this fight is to talk folks out of signing a union card, or into voting "no" in your union election. And they'll say just about anything to meet that goal. Bhavika Tekwani, one of the organizers on Mapbox's first union campaign, said that the company's leadership had one line that had an especially chilling effect: they told the staff that the union drive had caused the company to lose some financial backing.

> The day we announced, they basically said an investor withdrew. And it was very convenient. I personally did not believe it — or actually, I don't believe it went down exactly like that. But either way, they didn't need it to be true. They didn't need the timing to line up. But it just did. And it was really effective.

I want to underline what Tekwani said: "they didn't need it to be true." It's *possible* that an investor pulled out because of the union drive. But there is no way for the workers to verify that. It's a statement that has a whiff of truth about it — and that's what matters to management. It's enough to *hint* that union organizing is jeopardizing the company.

And hint your bosses will. They might suggest that a union would make the company less productive, or that employees are being "harassed" by union organizers; they might even suggest that the union campaign is "hurting the brand," and discouraging job seekers from applying to the company. At Mapbox, leaders took to the company Slack to accuse union organizers of bigotry and xenophobia, all because the union wanted assurances that jobs wouldn't be relocated to other countries (https://ydatu.com/04-19).

I'll say it again: statements like these don't need to be true. In fact, they probably aren't true. They're a tactic, plain and simple — a tactic that's designed to exploit any nervousness or anxiety that members of your bargaining unit might be feeling. If your company's leadership can make people feel hesitant about supporting the union, they will. Because that hesitation will, in turn, make you and your coworkers less likely to take collective action together.

Always remember: *your bosses can lie to you.* You have legal protections under the NLRA to form a union. But there is absolutely nothing that legally requires your management to tell you the truth.

Retaliation

It's possible that your management might take their fight directly to individual employees. Clarissa Redwine, an early organizer of Kickstarter's union, talked to me about the retaliation she witnessed at Kickstarter after employees pushed back on a questionable product decision — one that went against the company's stated values. When coworkers spoke up, Redwine said, retaliation from execs was swift: "Management took people into rooms one by one and said, 'What you did was not all right; this is going into your record.'" The company began weaponizing performance reviews, using them to discourage or punish organizers.

And all of this happened *before* Kickstarter's workers decided to unionize. But Redwine said the leadership's retaliation definitely changed that. "Everyone was like, 'Whoa, okay: They can fire us for speaking our minds? Or just doing what's right for the platform?' And that's when the union drive really started." Redwine herself received glowing reviews at Kickstarter — *until* she became a union organizer. From that point on, she was accused of "failing to build trust with management" and of "not being a team player." Redwine was fired soon afterward, alongside two other organizers from the union.

Firing isn't the only way bosses might retaliate. A few organizers I spoke with told me stories of being denied promotions they'd previously been lined up for, or of being denied a job elsewhere in the company after being encouraged to apply. If the goal is to discourage pro-union workers or chill support for the union, your management has several tools they can draw upon.

At the end of the day, management has exactly one pro-worker response to a union campaign going public: to voluntarily recognize the union. The alternative? Pushing back on the need for a union, dismissing workers' concerns, or starting a long, bitter campaign of attrition against the people working the hardest for a better relationship with their work. If your bosses decide to suit up for a battle against their own employees? Well. That seems like a pretty good indication that your union was needed in the first place.

Your election

If your base of support holds throughout management's union-busting campaign, then the day of your election should be fairly calm. The election itself should be straightforward: it will be overseen by agents from the NLRB, and it may be conducted in-person, over mail, or some combination of the two.

In the run-up to the election, you might still be trying to shore up your support: talking to folks who are on the fence, and trying to convert them to *yes* votes. But it's also a good idea to check in with known pro-union members, and make sure they *remember* to vote. This outreach matters for a few reasons:

1. Voter turnout is terribly important. Your company's management will be working to make sure "no" votes show up to vote, so you need to ensure every "yes" vote knows when and how to cast their ballot. If you've ever done phone outreach for political elections, this might feel similar: when you reach out to someone, ask if they have a plan for election day; ask them when they're planning on voting. And if they have any questions, try to answer them clearly and openly.
2. The stakes for your election are, frankly, high. If your union loses its election, it can't file for another election for an entire year.
3. Winning your election by a wide margin can be a real boon for contract negotiations, giving your union — and its mem-

bers — a massive morale boost as you begin the bargaining process.

But hopefully the election will be both uneventful and calm, and you'll win the vote decisively. From there, your union will finally be recognized. Reaching this moment is, to be clear, quite the milestone. No, wait — it's more than that. It is an *utterly wonderful thing*. It's the end of an organizing journey that took you and your fellow union members months — if not longer — of hard, emotional work. When you get here, I hope you and your coworkers will celebrate the hell out of it. Because now that you've cleared that hurdle, it's time to begin the collective bargaining process.

STEP 4: NEGOTIATING YOUR CONTRACT

I'll start here: winning your union election will immediately improve things for you and your coworkers. Your employer is required to maintain the status quo at work, and cannot change your working conditions without negotiating with the union. Additionally, all members of the bargaining unit gain access to a set of powerful legal protections known as *Weingarten rights* — essentially, allowing you to request that a union representative be present during any situations that could result in disciplinary action being brought against you (https://ydatu. com/04-20).

But more broadly, the period following your election is a time of transition. You'll shift from working with your union organizer, to planning bargaining sessions with new faces from your union's local branch. Your fight will be changing its focus, too: you'll be shifting from building union support over furtive Zoom calls and Signal chats, to fighting at a bargaining table for your union's contract proposals. You're going to wind down your organizing committee and stand up a new committee — one focused on winning contract negotiations rather than counting union cards. This process will likely be much longer than the organizing phase.

But remember: you'll be doing all of this alongside your union. You won't have to do any of it alone.

Getting local

Once your union has been certified, you'll be assigned to a union local. As I mentioned earlier, this is basically a "branch" of your national union. (And in many countries outside the United States, they're often called just that: *branches*.) A union local's name is usually numbered: for example, "CWA Local 1051" is a branch of—you guessed it—CWA. The number doesn't really mean anything; it's just a unique signifier that gives members of one local an easy way to distinguish themselves from members of another local.

Your union local will usually be the one closest to your company's headquarters. That doesn't necessarily mean it's in the same city, or even the same state—every national union has a different approach to drawing its maps. But once you're paired up, your union members will *also* be members of that local. And much like your own union, that union local is a democratic body with its own elections, elected officers, and bylaws. It's worth familiarizing yourself with them when you have the time.

But just as importantly, your union local will pair you up with a *lead negotiator*. This person might be a negotiation specialist who works directly for the union, or they might be a lawyer hired by the union to lead negotiations. Either way, they'll show up with experience and strategies for negotiating collective bargaining agreements, and they'll be responsible for representing your union in your upcoming bargaining sessions. If you have questions for them about the process, ask them: they've done this more than a few times, and they're here to help your union win.

Forming your bargaining committee

Before negotiations start in earnest, your union will form a *bargaining committee*. These are members of the unit who are interested in representing your union at the bargaining table. The number of people on your committee can vary, depending on the size of your bargaining unit; most unions I interviewed had elected three bargaining committee members, but a few unions had four. (One larger union had *eight*.)

Being on the bargaining committee is a lot of work. In fact, the decision to participate in bargaining gave pause to a few organizers I interviewed. And that is perfectly understandable! Many organizers are *fried* by the time they've reached the end of their organizing campaigns. Since being part of the bargaining committee means you'll attend all the negotiating sessions, and will advocate for the contract proposals your union is putting forward, it's worth thinking long and hard about whether you've got fuel in the tank for what's next. If you do, that is fantastic news—because it means you'll have the opportunity to represent your fellow union members and fight for a contract that could lead to better lives for all of you.

Sitting at the bargaining table

The bargaining sessions themselves have a fairly regular structure. You'll convene on a set schedule that's mutually convenient to both sides—some unions told me they were meeting every other week; one mentioned they had adopted a three-week cadence. At the sessions, two parties sit across from each other at the negotiating table:

- On the worker side of the table, there's your union's bargaining committee, as well as your lead negotiator.
- On the management side of the table is a small committee of managers or leaders, as well as their legal representation. Just as with your lead negotiator, their lawyers might come from your company's existing in-house counsel; frequently, though, they're hired from firms who specialize in negotiating labor contracts. (It's even possible they'll be the same lawyers who led your employers' union-busting campaign.)

In each session, one side will present a proposal for terms they'd like included in the contract—some language related to salary bands, say, or to paid time off policies. They'll explain the proposal and their rationale for including it in the collective bargaining agreement. A few questions might be asked, but there's usually very little open debate or argument: instead, both sides will decamp to review the proposal, and the session ends. By the time the *next* session rolls around, the other side may have a counterproposal to offer, or possibly a proposal of their own. If they do, they'll introduce it, argue for why it has to be included, and then both sides decamp again to discuss.

And that's the cadence of collective bargaining. Management and your union hammer out the language of the contract session after session — paragraph by paragraph, line by line — until there aren't any objections left. At that point it's considered a *tentative agreement*, or TA, one that's been provisionally approved by both sides of the bargaining table. From there, it goes off to your entire bargaining unit, which votes on whether to ratify it.

Now, how do you figure out *what* you're proposing? Well, that's going to be driven by your membership. For some unions, that begins with a data-driven approach. Avital Baral, a member of ActBlue's Tech Workers Union, said: "We basically designed a bargaining survey and asked people to rank their priorities, so that the people going into negotiations can be really informed about what people ultimately want." Mike Murray, who was part of the bargaining committee at EveryAction, said they took a similar approach: "We sent out a survey to everyone because we wanted to know, *What are your highest priorities?* Is it wages? Is it healthcare? Is it job security?" But Murray added that in addition to the survey, EveryAction's bargaining committee *also* conducted a set of listening sessions for everyone in the bargaining unit. He said that the combination — in-depth discussion, paired with high-level survey responses — really helped him better understand what their members needed.

In short, your bargaining committee should work to have as much input as possible from the broader bargaining unit, to make sure they're accurately representing everyone's shared priorities. But the tools they use to *gather* that input — online surveys, one-on-one conversations, status update meetings, collaboratively edited documents — can vary from union to union.

Structuring the sessions

Now, you and your lead negotiator can decide *how* you structure your negotiation sessions. Generally, there are two different models for running them:

1. First, there are *closed bargaining sessions*, where the only people who attend are the people sitting at the negotiating table: your company's representatives, and your union. There aren't any outside observers. Between sessions, your bargaining committee may provide status updates to the rest of your unit on what has been discussed. (Or it may not.) But otherwise, the door to the negotiating session is *closed* to outside observers. They'll see the draft contract when it's time to vote on it.

2. By contrast, there are *open bargaining sessions*, which are open to everyone in your bargaining unit. Not everyone will actively participate in the session; unless some plan has been made beforehand, most people will simply be present in sessions as silent attendees. That means they'll have access to every bargaining session and will be able to watch individual contract proposals (and counterproposals) get presented.

The decision to adopt open or closed bargaining sessions is a strategic one — one best left up to your bargaining committee and your lead negotiator. And it's worth noting that your negotiator may have a strong preference here. I spoke to workers at two different unions whose respective lead negotiators flatly stated: "I don't do open bargaining." Now, from what I gathered, these negotiators had some sound reasoning behind their refusal. One is simple logistics: if you're trying to make open bargaining sessions open to the largest number of people, that may mean you'll need to schedule them outside the workday. But I suspect some of their reasoning was stylistic. These negotiators had been fighting — and winning! — contracts for years, using a tried-and-true method. So why break what works?

Well, a growing body of evidence and testimony suggests that open bargaining sessions don't just result in stronger contracts; they result in stronger unions, too. A 2021 report from the UC Berkeley Labor Center stated that open bargaining has significant benefits, even just by "having every worker show up at negotiations at least once."

> The alternative [to closed bargaining] is a collective-negotiations process that invites, if not directly engages, the entire unionized workforce. In selecting our cases [for this report], we required radical transparency as the starting point for the negotiations process: this foundational practice can ultimately transform a union and lead to greater overall worker participation in the life of the organization. (https://ydatu.com/04-21)

Contract negotiations are, literally and figuratively, a fight. During bargaining sessions, the union will fight for issues that improve the lives of workers at the company: proposals like more regular hours, better pay, clearer paths to promotion, or more equitable review processes. And management will fight against proposals they see as too "costly."

That's why transparent, open bargaining sessions can be so powerful. When they attend those sessions, your union members will *see* that oppositional dynamic in action. I spoke with labor researcher Emily Mazo, who agrees. In fact, she told me that management's behavior in bargaining sessions can be *wonderfully* helpful in galvanizing your union. "Management can be the best inoculation tool," she told me. "I would hire a skywriter to write 'Have open bargaining.' It's the most important thing that workers can do."

In other words, when management fights against your union's demands, it sends a clear signal to each of the attendees — namely, that your company cares more about preserving the status quo than improving the lives of its workers. And that's not going to go unnoticed by any union members in attendance. That is, in fact, going to be *extremely* noticed.

Just as importantly, it's going to send another signal to employees: that their union is *fighting* for them. Your bargaining committee gets to be seen when it pushes back on a spiteful counterproposal from management, or when it calls out a talking point from your employer's anti-union lawyer.

Effectively, the bargaining sessions become a means to energize your union. They can demonstrate in real time the *value* of your union to its membership, and invigorate those members for the fights ahead. At least two tech unions — Code for America Workers United and the New York Times Tech Guild — have found this to be true, as they're using open bargaining sessions to negotiate their first contract.

But regardless of whether your bargaining sessions are open or closed, it's critical to keep your union as engaged and informed as possible. Because once you have a tentative agreement in place, your entire bargaining unit has to vote on it. If they approve it, then your agreement is ratified, and it becomes your final collective bargaining agreement — your union's first ratified contract.

One big union

I want to take a moment to underline that *if* in "if they approve it." Your bargaining committee will do its level best to fight for a contract that represents the interests and concerns of the bargaining unit. But concessions are inevitably going to happen. Decisions will be made during negotiations that will shape the CBA's final language, which will, in turn, impact your union.

How, then, do you keep the rest of your unit informed, so they're not surprised by what's in the contract when they're asked to vote on it? For most unions I interviewed, email newsletters were a big help. Between sessions, the bargaining committee sent digests out to the membership, updating everyone on the latest developments: what had been discussed in the most recent session, what was planned for the next one, and any items that needed action or research.

But it's important that you're not just broadcasting. After all, the point of building a union is to build collective power *as a union* — and not just for those making decisions at the bargaining table. In fact, one quiet thread that ran throughout my interviews was that participating in negotiating sessions was, in its way, a privilege. Bargaining, much like organizing, involves a considerable amount of work, and that often means the process can easily exclude parents, or those who are unable to work long hours, or colleagues who work a second job because their paycheck isn't enough — in other words, the people who most need to be heard when addressing workplace harms.

So how *do* you keep the rest of the unit not just informed, but as engaged as possible throughout the bargaining process? This is, I think, another point in open bargaining's favor. But regardless of how transparent your bargaining sessions are, ensuring your union remains mobilized will require a concerted effort.

When America's Test Kitchen United reached the bargaining phase, they set up a mobilizing committee — or MobComm for short. (Full disclosure: when I heard that nickname, I clapped for joy on the call. Yes, I'm fun at parties.) MobComm is essentially a Signal chat filled with leaders from each department across the company, each of them motivated, pro-union workers. What's more, they were recognized *as* leaders in the respective teams and departments they worked in. They each agreed to act as a hub of communication and — you guessed it — *mobilization* for their respective parts of the organization. If there was a question the bargaining committee wanted to pose to the membership, or something they needed the rest of the unit to provide feedback on, they could turn to MobComm for support.

The NYT Tech Guild took a more structural tack when writing its contract proposals: it set up a series of "article committees" staffed by members of the bargaining unit. Each functioned a bit like a working group focused on a specific topic area: compensation, healthcare, and so on. Vicki Crosson, a member of the Tech Guild, told me that every article committee was initially tasked to be bold with their thinking: "We said, 'Pick your gold standard, and just *run* in that direction.'"

Each article committee used a number of tools to approach its work: reviewing language from other unions' ratified contracts, surveying members, or simply talking with coworkers. (Or all three.) Their recommendations were then translated into contract-friendly language and used to inform the bargaining sessions. For me, as impressive as this process is from a democratic standpoint, the other benefit is that it kept their *very large* membership engaged. Here's Crosson again:

We've got one hundred people [involved in the article committees] out of the six hundred-ish people in our unit. And there are some people involved in multiple committees. I think it's really inspiring to see that many people who want to be at least aware of what's going on: they want access to the proposal docs, and they want to be able to leave comments.

In many ways, the bargaining phase involves a different kind of organizing campaign, one that's happening *while* you're negotiating your contract. Tackling that problem in the most effective *and* equitable way possible will help your union build support for your contract, sure — but you'll be building power that extends far beyond its ratification, too.

WHAT HAPPENS NEXT

MJ Flott was involved in organizing the union at EveryAction, and spoke to me shortly after their membership voted to finalize its first CBA. And Flott was incredibly enthusiastic about what their union had won. "I have a coworker who's been living with her parents because she can't afford her own place to live," they told me. "And I got to tell her that she's getting a ten thousand dollar raise." Flott went on: "The bargaining process can leave people frustrated, or worrying we didn't get enough. But for some people, this is life-changing amounts of money. It is *fucking awesome* to tell people they're getting this from our contract."

For many, a ratified contract means the value of having a union isn't just apparent — it's *established*. Ben Harnett, a member of the New York Times Tech Guild, told me:

You get a lot of benefits from having a union, even before you get the contract. You have solidarity; you have Weingarten rights. But it really is the power of winning the contract that shows what a union can achieve. And once you've shown people that you can actually win tangible workplace benefits and changes, that's the point at which the union becomes sustainable.

But at the end of the day, it's not *just* about proving value. Many organizers felt that they'd truly contributed to something fundamentally different at their work—something *good*. Dave Stern, a data scientist in ActBlue's Technical Workers Union, said he felt gratified about what he and his coworkers had accomplished in getting their union recognized: "It's, like, knowing that if I'm not here forever? I'll have left it better than I found it. If you join a nonunionized workplace, and you leave it unionized, you should feel good."

Building a union, ratifying a contract together—these aren't small things. For many tech workers I spoke with, the experience quite literally changed their lives. For those with ratified contracts, their unions enabled them to win real, material gains at work: these agreements close up pay disparities, improve workplace safety, and fight for more equitable benefits. The contracts being won are fundamentally changing the nature of tech work.

But today's labor movement is changing *tech workers*, too. Several people told me that the experience of forming a union helped them realize labor organizing was something of a calling for them. When taking a new job, some workers decided to help organize a union at their new company; other workers decided to leave their jobs and pursue work as full-time organizers. And that is a very real and profound kind of power, too.

In the next chapter, we'll talk about what else our power can—and should—be used for.

5 TECH SOLIDARITY, FOREVER

> "The voices of the powerless are not usually heard in technological deliberations, and we have no civic equivalent of the family's practice of "one cuts, the other chooses."
> —**Ursula Franklin,** *The Real World of Technology*

IN 2017, THE PRODUCT TEAM at Airbnb built a proof of concept for a new prototyping tool, one in which a user interface could be built simply by sketching on paper (**FIG 5.1**). If you watch the accompanying video, you'll see that the designer draws some rough shapes on a piece of paper, and then slides that paper under a camera. From there, a computer analyzes the picture, identifies the shapes, and associates them with specific UI components from Airbnb's design system. Equipped with that information, the computer renders a finished, web-ready prototype in the browser.

When Airbnb shared this prototype, the team said that their work was shaped by this guiding principle:

The time required to test an idea should be zero.

FIG 5.1: From rough paper sketch to finished prototype, all in a few short seconds (https://ydatu.com/05-01).

There's a lot that's appealing about this idea. As a mission statement, it's a fine one. But it's worth remembering that *people* — you, me, all of us — live and work in the hours, minutes, and seconds we're trying to streamline out of our product cycles. And our industry has a terrible track record of asking about the second-order effects of its product decisions. For example, if we *could* instantly create designs in a browser, what happens next? What jobs would be changed as a result? Who would be impacted?

In the years since Airbnb's experiment, it seems we've been asking those questions more and more often. In fact, there's been an explosion of "generative artificial intelligence" software platforms, all of them focused on automating tasks that've traditionally been the domain of engineers, writers, or designers — all of them released in rapid succession since 2021. GitHub released Copilot, a utility that can parse plain-language inputs from a user, and suggest working code snippets that fulfill the desired request. The research laboratory OpenAI released ChatGPT, a tool that allows users to ask questions in plain English, and then returns natural-sounding answers. And there are AI art tools like DALL-E, Midjourney, and Stable Dif-

FIG 5.2: An image created in Midjourney, using the prompt "a mechanical dove" (https://ydatu.com/05-02).

fusion, which can generate stunning images in response to simple text prompts (**FIG 5.2**).

These utilities are remarkable, full stop. By simply typing a few phrases into a chat interface, their users can quickly produce something that resembles code, content, or art. And frankly, I'm awed by how rapidly these tools are evolving. In 2022, AI image generators could barely render a recognizable human face; a few months later, they were producing high-resolution, photorealistic pieces. In March 2023, OpenAI promoted the release of GPT-4, the latest version of its large language model (LLM). As part of the publicity around the launch, they released a video showing GPT-4 generating a simple web page by analyzing a rough, hand-drawn sketch (**FIG 5.3**). These tools are evolving remarkably quickly. It's hard not to look at their current state and see today's shortcomings as their floor.

These utilities aren't remotely perfect, mind you. Nor are they free from controversy. They're able to produce such stunning results because they've sifted through staggering amounts of training data—of preexisting code, content, or art—often without regard for whether the original work's license or copyright permitted it. And *what* they produce often has significant flaws: on its website, ChatGPT notes that it "may occasionally

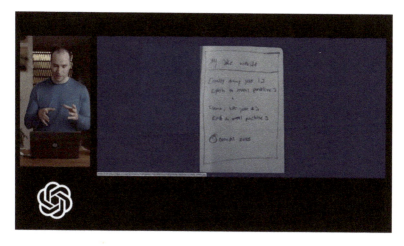

FIG 5.3: GPT-4 analyzes a simple-looking sketch, and uses it to build a rudimentary page with HTML and CSS (https://ydatu.com/05-03, video).

generate incorrect information" — and indeed, the prose it spits out often sounds realistic, but isn't factually correct.

But putting these shortcomings aside for a minute, I have to admit I'm impressed and terrified by these tools in equal measure. As a designer, I can't help but wonder what they mean for the nature of my work. And maybe you're curious, too: after all, when everyone has access to tools like these, what *does* it mean to be a "designer," an "engineer," or a "writer"?

THE MACHINE COMES FOR ME, AND FOR THEE

Again, these tools *are* inspiring. From an engineering perspective, they're brilliant; on a personal level, I've always dreamed of something like this, where software could instantly translate my ideas into a finished, polished product. But they exist in an industry that's deeply invested in the idea that, as the Airbnb team said, the time required to test an idea should be zero — and that the number of *people* involved should be close to zero, too.

I'm reminded of a blog post by Dave Rupert, in which he shared some early impressions of Copilot, GitHub's code generation tool. He wrote about how its code suggestions often worked, but weren't always what he was looking for. However, the net effect was that the software changed how he approached his work:

> My biggest adjustment with using Copilot was that instead of writing code, my posture shifted to reviewing code. Rather than a free form solo-coding session I was now in a pair-programming session with myself (ironically) in the copilot seat reviewing. [...] The end goal of programming is working software and the robot can suggest code faster than I can write code. (https://ydatu.com/05-04)

According to Rupert, the primary responsibility for creating code shifted away from him and to the software. As a result, he found himself adopting more of a supervisory role, spending his time reviewing the code that the tool produced. (And occasionally, rewriting it.)

There's a term for this in economics: *deskilling*. As a new technology is introduced to an industry, it may be able to complete tasks that had traditionally been performed by human workers. And as the technology matures, it can be overseen by fewer workers, which gradually lowers wages and may eliminate jobs. In other words, the demand for skilled labor is gradually reduced — or eliminated altogether. Once the technology reaches a certain critical mass, workers are moved into roles they would traditionally have been seen as overqualified for, while less skilled workers are driven from the market altogether. Every worker is, quite literally, *de*-skilled by automation.

We're starting to see examples of this happening already, where workers are losing jobs to this new crop of AI-based tools. Jason Colavito, a writer, complained on Twitter that he lost a copywriting client to a piece of software:

> A client informed me that he will no longer pay me to write content for his website because A.I. can write it for free, but he wants to pay me a fraction of my usual rate to "rewrite

it" in different words so it can pass Google's A.I. detection screening. (https://ydatu.com/05-05)

And in May 2023, it was revealed that BuzzFeed had published over forty "AI-generated" travel guides:

> Right now, there are 44 posts covering destinations like Morocco, Stockholm, and Cape May, New Jersey. The articles are "written with the help of Buzzy the Robot (aka our Creative AI Assistant) but powered by human ideas," Buzz-Feed says on Buzzy's profile. The top of each story I've seen includes a line noting that an article was "collaboratively written" by a human and Buzzy. (https://ydatu.com/05-06)

Since being purchased by a new corporate owner in 2020, the editorial staff at CNET has been gradually emptied out, and much of the site's content is now being generated by an AI utility:

> Internally, there has been unease among CNET staff at their corporate owners' use of artificial intelligence — though staff was assured the current test is limited in scope. But layoffs and restructuring, coupled with the lack of clarity on the use of new tools, are causing some to worry about what the creep of AI signals for the venerated site so many journalists were drawn to.
>
> After multiple rounds of layoffs last year, dozens of people lost their jobs, from audience and copy teams to CNET cars staff. Entire teams were decimated, one former staffer says, and people continue to leave "in droves," fearing more layoffs are around the corner. (https://ydatu.com/05-07)

When the Writers Guild of America (WGA) voted to strike in 2023, it was because their bargaining sessions had broken down, at least in part, over generative artificial intelligence. Specifically, the WGA had been pushing for language in their minimum basic agreement (MBA) — effectively, the union's collective bargaining agreement — that would:

> Regulate use of artificial intelligence on MBA-covered proj-
> ects: AI can't write or rewrite literary material; can't be used
> as source material; and MBA-covered material can't be used
> to train AI. (https://ydatu.com/05-08)

In other words, the union was seeking contractual guarantees that studios wouldn't use AI to write or edit scripts, effectively reducing the need to hire writers. The entertainment industry's response? They rejected the union's proposals, offering instead to hold "annual meetings to discuss advancements in technology" (https://ydatu.com/05-08). Seems like the WGA was right to be concerned.

And we should be, too.

We're far from the first industry to see the nature of our work change, and so quickly. I spoke with Keith Hogarty, a longtime union organizer with CWA. Hogarty has been organizing for twenty years, helping workers from various industries form unions. And he's worked in the telecommunications industry for decades.

During our chat, Hogarty told me how he managed to get work installing residential networks during the internet's early days. The work was, he remembered fondly, *really good*. "My job was a cushy job," he remembered. "We didn't have enough people to do the work: to install jacks for people who needed internet in their houses. Then," he added, "the smartphone came out. Pretty quickly, my cushy job wasn't cushy anymore. I was in a fight for survival."

I think about Hogarty's story often. Over the span of a few short years, the easy work of installing fiber-optic networks — working indoors, and getting paid well to do it — quickly became scarce, thanks to the deployment of cellular networks. The easy work dries up; the value of your labor becomes lower over time.

None of this happens by accident. Technically speaking, software alone can't displace workers — rather, it takes time, effort, and concerted investment from those who *want* to deskill workers. And as it happens, there is no small amount of capital invested in AI-backed automation: Stable Diffusion's parent company received over $100 million in venture capital funding at the end of 2022 (https://ydatu.com/05-09, requires registration); OpenAI, the company behind DALL-E and ChatGPT, has a $29 billion valuation (https://ydatu.com/05-10).

But money alone isn't enough to introduce wide-scale automation to an industry; it also takes influence. And right now, many of tech's most powerful investors are talking openly about how human labor is dramatically overvalued in the tech industry — that there are simply *too many tech workers*. In no particular order:

- Marc Andreessen, the noted venture capitalist, recently opined that "the good big companies are overstaffed by 2x. The bad big companies are overstaffed by 4x or more" (https://ydatu.com/05-11).
- Nat Friedman, the investor and former CEO of GitHub, says on his homepage that he believes "many tech companies are 2-10x overstaffed" (https://ydatu.com/05-12).
- A venture capitalist named Brad Gerstner wrote to Mark Zuckerberg, demanding that Meta reduce its head count. As Gerstner put it, "It is a poorly kept secret in Silicon Valley that companies ranging from Google to Meta to Twitter to Uber could achieve similar levels of revenue with far fewer people" (https://ydatu.com/05-13).
- A billionaire investor named Christopher Hohn wrote a similar letter to Alphabet's CEO, demanding the company take "aggressive action" since "the company has too many employees and the cost per employee is too high" (https://ydatu.com/05-14, PDF). After Alphabet conducted a massive round of layoffs, Hohn wrote again to demand Google reduce its head count even more — by another 20 percent (https://ydatu.com/05-15, PDF).

I pause here to note that Hohn is reported to have made nearly one million pounds *per day* in 2021 (https://ydatu.com/05-16).

You don't have to look far to find more examples of this refrain, sung by voices that are almost entirely male, predominantly white, and very, very, very, very, *very* wealthy. As the song goes, the tech industry is overstaffed and bloated, and your company needs to fire anywhere from 50 to 75 percent of its employees to reduce its costs. The work *you* do, and the work of your coworkers, is disposable: a rounding error on the books, remedied by sweeping layoffs. I feel the need to disclose that I removed several swears from this fucking paragraph while typing it.

(Whoops, sorry. Missed one.)

Ahem. It's possible that folks who hold to this particular line of thinking—that a company can somehow improve its performance after shedding more than half its staff—are watching the industry's mass layoffs with interest, and waiting to see whether or not their theory will play out. For example, how well can Twitter function after firing most of its workers? Can it, as Brad Gerstner argued, "achieve similar levels of revenue with far fewer people"?

At least in the months following Twitter's mass firings, the answer seems to be: "lol no." Users have reported numerous features failing unpredictably, with Twitter's app and website occasionally becoming inaccessible in different countries. Hate speech, misinformation, and political disinformation ads have crawled over the platform—which is also unsurprising, given that Twitter's moderation and safety teams were hollowed out by the layoffs (https://ydatu.com/05-17, subscription required). As a result, advertisers have abandoned the company in droves: at the time of this writing, more than half of Twitter's top thousand advertisers have paused their campaigns on the platform. By December 2022, the company's revenue had cratered by 40 percent over the prior year (https://ydatu.com/05-18).

But frankly, I don't believe any of these investors really care. They're not watching to see how well the tech industry's layoffs are "working," because they're not angling for a better, more productive tech industry. They're reacting to rising interest rates, and to a stock market that soured dramatically on the tech industry in 2022 (FIG 5.4). This idea of mass layoffs as some kind of industry corrective is a wholly unserious idea, cooked up by a gaggle of prominent (and affluent) voices who are solely concerned with better returns on their investments. Wealthy investors calling for "fiscal discipline" in the guise of mass layoffs should have their motives questioned, loudly and publicly.

FIG 5.4: A screenshot of a Vanguard index fund, made up of stocks of various companies in the tech sector. The fund lost a third of its value in 2022.

But let's put their motives aside for a minute, because there's a larger point to be made here.

THIS, TOO, IS FOR EVERYONE

Historically, tech workers have resisted the idea of unionization because of the privilege we've enjoyed: for the most part, we're paid well, our benefits are good, and our work is often highly valued by the market. What protections could we possibly need?

But the privileges we *do* have as tech workers are fleeting — or, more specifically, they're actively being eroded. Thanks to technological advances and staggering amounts of capital, we have incredibly robust automation that's changing the nature of digital work. Additionally, we have an army of influential investors loudly arguing that tech workers are too numerous and dramatically overvalued. These are two mightily powerful forces, and they're both keenly interested in lowering the value of our labor. I mean, not *five days* after laying off ten thousand workers, Microsoft announced a $10 billion investment in OpenAI, the AI research organization behind ChatGPT and DALL-E (https://ydatu.com/01-12; https://ydatu.com/05-19, subscription required). If that doesn't seem like a signal about the kind of work that gets valued in our industry, I don't know what does.

And I don't think it's a coincidence that this is all happening amid a resurgence in our industry's labor movement. Tech workers have been organizing, and forming unions, and fighting for more power — power to define a more equitable vision of what tech work can be. And now we're getting told we're overvalued, and getting laid off by the hundreds of thousands.

This is a fight, and we're in it.

Make no mistake, we're going to win this fight. But to do so, we have to act while we enjoy the privileges we do have, and the power that comes with them. We have to start organizing our workplaces *today*, at a scale that's unprecedented for our industry. And that organizing starts with you: with the union that you and your coworkers are forming.

I realize this sounds a bit daunting. But it might be helpful to remember that we're not the first workers facing an investment class that wants to hollow out our workforce, and we're *far* from the first workers to grapple with the ravages of automation. That doesn't make the fight easier, of course. But studying history might help us better prepare for the road ahead.

Here's one example: the International Longshore and Warehouse Union (ILWU) represents thousands of dock and port workers on the West Coast of the United States. And in the 1950s, the shipping industry was making a concerted push to introduce *containerization*, a highly automated shipping process for loading cargo. Adopting the technology would replace most of the hand loading that had traditionally been done by unionized dock workers. But the ILWU realized that simply fighting containerization wasn't an option, because it could potentially shift work to ports in other countries that *did* adopt the new technology (https://ydatu.com/05-20).

Instead, they did something perfect: they fought for a contract that ensured workers would be paid *regardless of whether or not they worked*, thereby providing them with a guaranteed weekly income. In other words, the ILWU fought for a contract that ensured its workers wouldn't just be protected from the automation technology that threatened their livelihood — their union contract let these workers *profit* from automation.

The ILWU isn't the only union to have won this fight. Lisa Kresge, lead researcher at the UC Berkeley Labor Center, published a paper filled with examples of various unions' collective bargaining agreements, and how they were used to respond to technology (https://ydatu.com/05-21, PDF). In the paper, Kresge outlined three broad strategies for using contractual language to mitigate technological harm:

1. Mandate that the union be involved in the decision-making process behind adopting a new technology.
2. Create protections that protect workers from any technologies introduced to the workplace. (The ILWU contract is a landmark example of this.)
3. Define limits on *how* technology can be used in the workplace — for example, by protecting workers from surveillance and monitoring.

This is a brilliant and timely framework, one that neatly grounds the collective bargaining process in the fight against creeping automation. Our union contracts can protect us today, and in the fights to come. That's why it's critical that we start asking ourselves what we need in case the worst effects of automation impact our industry. What protections would we need? What *benefits* do we want to gain?

To answer those questions, we have some work to do: we have to organize our workplaces, and form unions, and fight for strong contracts.

And then we *keep* fighting.

But as we organize — and unionize, and *fight* — we have to realize that we're not alone in this struggle. In fact, we're organizing alongside workers who've been dealing with the worst of these conditions for years.

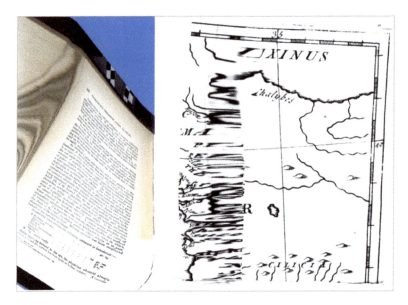

FIG 5.5: A collection of beautiful "errors" from Google Books: misaligned or distorted pages, some overlaid with digital artifacts (https://ydatu.com/05-22).

THE HIDDEN HANDS THAT MAKE TECH WORK

On that note, I have two pieces of art I'd like to recommend.

Let's start with Krissy Wilson's The Art of Google Books (https://ydatu.com/05-22), specifically the posts tagged #hand (https://ydatu.com/05-23). The first thing you might notice are the pages, and how *off* some of them look. Many of the scans are blurred, or misaligned, or covered with digitization artifacts (**FIG 5.5**). Some of them aren't even pages at all. And as you scroll through these images, you'll likely spot a hand or two, each gently holding a page in place for scanning (**FIG 5.6**).

But look at the hands — not the pages. These are the hands that hold pages in place for digitization, before they're made accessible online. These are the hands that, quite literally, built Google Books, Alphabet's online digital library.

Take a closer look at those hands.

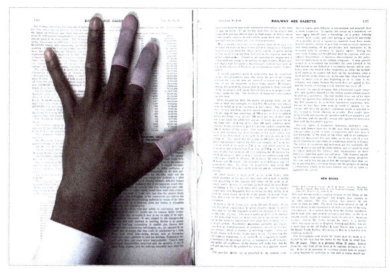

FIG 5.6: An entry from Krissy Wilson's The Art of Google Books" shows an accidental scan of a worker's hand (https://ydatu.com/05-24).

In 2022, Google's own corporate diversity annual report stated that its tech workforce was 62.4 percent male, and 37.6 percent female; Black+ and Latinx+ employees made up 5.3 percent and 6.9 percent of Google's workforce, respectively (https://ydatu.com/05-25). But nearly all of those hands—these *workers*' hands—have darker skin than I do; on many of the photos, you can spot a pop of color from nail polish, or a glint of light from the odd piece of jewelry. In many ways, the hands scanning Google Books feel like an inversion of the broader Google workforce: the hands in these pictures are overwhelmingly coded as feminine, and belong almost entirely to people of color.

Those hands belong to data entry workers, a throng of hourly contractors employed as part of Google's Scan Ops workforce. And that brings us to the second piece of art I'd like to recommend to you, Andrew Norman Wilson's "ScanOps" project (https://ydatu.com/05-26). Wilson learned about the scanning operation back in 2007, when he was working at Google's headquarters. You see, people employed by Google, whether salaried or not, were issued different-colored badges to communicate their status: full-time Google employees were issued white badges; contractors were issued red badges; interns were issued green badges. Wilson, a red badge-bearing contractor himself, had access to free meals, on-campus events, and transportation, even though he didn't have access to white-badge benefits, like stock options.

Wilson noticed that Google Books' data entry workers had *yellow* badges, and even less access to the benefits enjoyed by other workers — even other contractors at Google. The Scan Ops employees weren't allowed to eat free meals in the campus's many cafeterias, take free rides home on the corporate shuttles, or even ride free bikes around campus.

In 2011, Wilson created a short video titled "Workers Leaving the Googleplex" (FIG 5.7). In it, he recorded the Scan Ops team leaving their building at the end of a shift — 2:15 p.m. on the dot, after starting work at 4:00 a.m. It almost feels, as Wilson says in the video, "like a bell just rang, telling the workers to leave the factory."

According to the documentary, Google fired Wilson not just for recording this footage, but for speaking with some of the workers; Wilson maintains he was simply trying to understand more about the terms of their yellow-badge employment. In other words, Google allegedly fired Wilson for breaching the secrecy Google placed around a new, hidden class of contractors. Contractors who were digitizing content for Google's products, who were beginning their shifts well before dawn — all while they received less pay and fewer benefits than other tiers of workers at Google's headquarters.

But this practice extends far beyond Google. In fact, it's one that enjoys a long, rich history in our industry. The tech industry's gains have, time and again, been made precisely *because* of workers like Google's Scan Ops contractors: hourly workers performing rote, repetitive tasks, working long hours for little pay. And the reason this strategy has worked is that these workers — and their labor — have, whenever possible, been hidden from sight.

FIG 5.7: Andrew Norman Wilson's "Workers Leaving the Googleplex" shows digitization contractors streaming out of a building at the end of a long day (https://ydatu.com/05-27, video).

The hands that teach the machine

We've already looked at some of the advances our industry has made in "machine learning," which is often described as a type of artificial intelligence. But to put it simply, machine learning software is designed to become more accurate over time by itself — without human intervention. Think of your favorite streaming service's recommendation engine: any time you rate a movie or show, their software uses those inputs to "learn" how to provide you with better recommendations in the future. In other words, a machine learning algorithm will review the data it has processed in the past — its prior experience, if you like — to improve what it produces in the future. In other words, machine learning software is about *automation*: it improves its output without needing a human to program better results.

But these advances in machine learning are, at least in part, driven by something called "training data." Training data is fed into machine learning algorithms to make them more accurate. It quite literally is responsible for teaching them: working from training data, machine learning algorithms learn how to identify objects and relationships, and how to make decisions. Put another way, if you show an image recognition algorithm one million pictures that you definitively know contain cats, the algorithm will be able to more accurately identify — you guessed it — cats.

FIG 5.8: Humans must annotate "meaningful" pieces of training data, the way I've labeled our cats in this photograph.

And training data often comes from humans. Humans have to sift through all those photos not just to confirm that, yes, they're photos of cats, but to highlight the part of the photo that contains a cat, and to label that collection of pixels accordingly (FIG 5.8). That act of annotating and labeling — done by human hands and eyes — is what makes the data meaningful to a computer, which is how the algorithm begins to learn.

That annotation step applies to other kinds of training data, too. Whether it's audio recordings, text transcripts, or videos, those files are often reviewed and labeled by human annotators. Doing so improves the quality of the data, which improves the accuracy of the machine learning models that are subsequently *trained* on that data. In other words, a human does work, so that the machine can learn.

As the industry's attention has shifted to machine learning and artificial intelligence, entire industries have sprung up around the annotation of training data, often employing underpaid contractors in countries with few labor protections. And this model makes for incredibly big business. In January 2023, *Time* reported that OpenAI had hired Kenyan laborers and tasked them with "cleaning up" ChatGPT, their artificial intelligence chatbot. ChatGPT had been producing violent, sexist, and racist statements, and these workers were hired to tag and identify the problematic sentences. They were, effectively, training the software to be less bigoted. And the work was grueling:

> All of the four employees interviewed by TIME described being mentally scarred by the work. Although they were entitled to attend sessions with "wellness" counselors, all four said these sessions were unhelpful and rare due to high demands to be more productive at work. Two said they were only given the option to attend group sessions, and one said their requests to see counselors on a one-to-one basis instead were repeatedly denied by Sama management. (https://ydatu.com/05-28)

OpenAI has a $29 billion valuation; Sama, the contractor who hired the outsourced labor, was reportedly paid $12.50 an hour. The Kenyan laborers were paid less than $2 an hour.

Obviously, this is exploitation; these people are being exploited. And none of it happened by accident. By opening offices in impoverished communities, where wages are low and where labor protections are minimal, outsourcing companies like this one — and their corporate clients — can maximize their gains *and* minimize their costs.

And imagine how precarious these jobs are. As image and text recognition technology improves, will these contractors still have jobs in a few years' time? It's quite possible that the data work they're doing is being fed into automated solutions that will eventually replace them.

The hands that make the machine safe

All modern social media platforms struggle with content moderation. If you've ever wondered how you can get through most of a day online without seeing truly disturbing content, it's due to the efforts of thousands of moderators, many of whom work as hourly contractors. These moderators spend their shifts sorting through flagged posts, searching for anything that might violate the company's terms of service. To be blunt: this is terrible, stressful, traumatizing work. These people work each day to sift through disturbing posts, imagery, and video, sanitizing social media feeds before they reach our screens — all for frighteningly low pay.

And we're only just beginning to learn the real, human costs of social media moderation. Back in 2014, Adrian Chen interviewed moderators who worked as contractors for companies like YouTube, Facebook, and Twitter, many of whom burned out after a few months of reviewing the darkest content posted on the platforms (https://ydatu.com/05-29). More recently, the *Verge* published an exposé on the practices of contractors working in Arizona as social media moderators, many of whom develop PTSD-like symptoms, and who can't even be guaranteed access to a bathroom during their short, supervised breaks (https://ydatu.com/05-30).

At the time these stories were published, the median Facebook employee earned $240,000 annually in salary, bonuses, and stock options (https://ydatu.com/05-31). By contrast, the contractors working as content moderators in Arizona reportedly earned just $28,800 per year. In addition to the low pay, these workers have almost no job security. Quite the opposite, in fact: many of these moderators are tracked incessantly, and can be fired if they make more than a few moderation mistakes in a week.

Nearly every contract worker hidden at the edges of the tech industry struggles with this. It's not just about the financial precarity they live with, working as they do for low pay and poor benefits, but about how surveillance and hard, inflexible metrics govern every aspect of their jobs. When I spoke with Drew Brandt about their time working at a firm contracted to a giant tech company, they talked about how their performance was measured each week. I still remember how Brandt said their employer's metrics made them feel "pure anxiety." They added:

> It dictates everything about the job, because you know that they can fire you at any time if you slip below those metrics. Even if you are otherwise doing fantastic work, it almost is like they don't care. [...] It really just starts to consume your life and makes you feel like you're not a good employee. Or that you are somehow deficient.

Several contractors shared similar feelings with me: that as they work, they're keenly aware that the quality of their work is measured against metrics set by their employers. What's more, those metrics can be changed at any time, for any reason — and the changes can cause you to lose your job if you fail to meet those new numbers. It is, simply, dehumanizing.

The hands that make the machine go

Amazon is a company heavily invested in automation, at nearly every level of its business. By analyzing purchasing data, their systems can help forecast product demand or identify fraudulent purchases. Machine learning algorithms help tailor product recommendations and promotions to you as you browse. Amazon even launched a chain of cashierless Amazon Go stores, in which you purchase products by simply picking them up, and without ever interacting with a human — all thanks to an invisible network of in-store cameras and sensors.

But most of Amazon's automation investment has been in its warehouses, the fulfillment and sorting centers that power the company's impossibly fast delivery times. For the last decade, Amazon has deployed over a half million robotic drive units across Amazon's facilities, using the machinery to ferry inventory to human workers for sorting, packaging, and shipping (FIG 5.9).

Although there's a considerable human element in this highly automated workflow, Amazon is doing its level best to automate even more of it. At the end of 2022, the company announced it was considering introducing a small but sweeping change in its warehouses: eliminating the humble-but-ubiqui-

FIG 5.9: An Amazon Robotics Drive Unit carries heavy shelves of inventory across a warehouse floor (Photograph from Amazon Press Center).

tous barcode, and replacing it with a new scanning process. Here's an excerpt from the press release:

> On a conveyor belt, the lighting and the speed of the item are relatively controlled and constant. If a person is picking up an item, there are a lot more variables to performing identification in-hand. The employee's hand might make item detection more challenging depending on how they hold it. In addition, if an item is being passed from someone's left hand to the right, it has to be identified faster. Robotics researchers are working to address these challenges. (https://ydatu. com/05-32)

I first learned about Amazon's announcement from the writer Tim Carmody, who covered the announcement in his (excellent!) newsletter, *Amazon Chronicles*. In it, Carmody suggests that the language Amazon uses here marks a shift from how it used to talk about its workers. And a significant shift, at that:

> [I]f someone at Amazon were writing this press release ten years ago, or maybe even two years ago, there would be some ritual genuflection about how this new technology is actually better for Amazon's warehouse workforce, reducing their physical and cognitive load, lowering their risk of injuries, and freeing them up for tasks better suited for humans to do.
>
> Amazon's press release has none of that. Human hands in the warehouse handling goods in this account can only be the problem, never part of the solution. (https://ydatu.com/05-33)

To hear Amazon tell it, their fulfillment processes are entirely too reliant on human workers and their slow, error-prone minds and hands. The solution, therefore, is simple: just replace those people with a proprietary — and automated — scanning process.

In the meantime, the company's doing its level best to treat its human workforce like the robots they work with. Nearly every aspect of an Amazon warehouse worker's day is tracked, measured, and quantified. There are cameras everywhere, always recording. Workers handle packages using handheld scanners, which Amazon uses to gauge how quickly inventory is moving toward its next destination. The company also gathers data *about its workers* from the scanners, and logs their "idle" time — which could include going to the bathroom, repairing a broken machine, or simply talking with a coworker — as a metric called *time off task* (TOT). If a worker accrues too much TOT, they can be disciplined by a manager, or even fired.

An especially moving anecdote about this appeared in the *New York Times*:

> Dayana Santos, 32, who started at Amazon's fulfillment center in New York City in June 2019, appreciated the metrics. "How can I do my job efficiently if the next person isn't doing theirs?" asked Ms. Santos, who sometimes raced with colleagues for fun. "Why does everything have to be a competition with you, Santos?" her boss would tease.

After months of praise from her managers, Ms. Santos had one very bad day. She had been working in robotics, but because her bus was late, she was sent to picking. She was offered a different assignment after lunch, but it never came through, and her station in picking was occupied. She traversed the warehouse looking for another one, racking up more time off task. That afternoon, she was stunned to discover that she was being fired. (https://ydatu.com/05-34, subscription required)

It's important to situate all of this against the terrible, frightful conditions of simply working in Amazon's warehouses. In addition to the everyday surveillance workers face, Amazon has been accused of enforcing other intolerable conditions at its warehouses, like eleven-hour workdays and timed toilet breaks. Workers are often required to work mandatory overtime shifts, depending on company promotions or holiday schedules. In the winter of 2016, Scottish news outlet the *Courier* reported that Amazon fulfillment center workers in Scotland had taken to sleeping in tents near the facility because they couldn't afford the commute to work (https://ydatu.com/05-35).

The result is that the physical demands on these workers can be grueling, and sometimes deadly. In 2020, Reveal and PBS NewsHour found that Amazon's warehouse workers were likely to get seriously hurt — injured to the point of being unable to do their jobs — twice as often as warehouse workers in other industries (https://ydatu.com/05-36).

Here are some examples:

- In 2017, Devan Michael Shoemaker and Phillip Terry were both killed in different Amazon warehouses: Shoemaker was run over by a truck in a Pennsylvania warehouse; Terry was crushed by a forklift in an Indiana warehouse (https://ydatu.com/05-37).
- In 2018, Cynthia Dixon effectively destroyed her back after just two months of lifting and scanning orders at an Amazon warehouse (https://ydatu.com/05-38).
- In the spring of 2020, Alberto Castillo contracted COVID-19 during a mandatory overtime shift at JFK8, the Amazon ful-

fillment center in New York City. In December of that year, after months of fevers that left him with extensive brain damage, he was brought home for hospice care (https:// ydatu.com/05-34, subscription required).

- In 2022, Rafael Reynaldo Mota Frias died while working a summer shift at an Amazon fulfillment center in New Jersey. He was forty-two, and a father (https://ydatu.com/05-39).

These names are only a few picked from a long, long list of people who've dealt with the dangerous, and occasionally deadly, working conditions at Amazon — conditions that workplace safety regulators and journalists alike have been sounding the alarm on for years. The costs of two-day delivery are frightfully high, and they're paid by Amazon's warehouse workers.

WE ARE, ALL OF US, TECH WORKERS

There's an essay by Mandy Brown that I'd recommend to you, in which she digs into the industry's reliance on the words "technical" and "non-technical" to describe certain roles. And she gets to the heart of why that distinction is so incredibly broken:

> People in "non-technical" roles are typically paid less, afforded less flexibility, and granted less prestige. The categories of "technical" and "non-technical" serve wholly to privilege those in the former, at the expense of the latter. But literally no product organization would survive a week without the deep — and, I'd argue, technical — expertise of the people who are usually lumped into the "non" bucket. (https://ydatu.com/05-40)

Brown's essay is worth reading in full. In it, she gets at the implicit and explicit values that our industry assigns to "technical" roles, and how that value assignment is used to perpetuate certain inequalities. Treating a certain kind of work — here, the "technical" roles tasked with writing code — as *more valuable* than other kinds of work assigns a certain value not just to the work, but to the people who do it.

We can easily see how that language translates into real-world harms. If the "technical" roles in your organization are dominated by members of a certain demographic, and those roles are typically paid (and respected) more than "non-technical" roles, then what systems of disenfranchisement are upheld by that distinction between technical and non-technical workers? In other words, whose work is valued and respected, and whose work isn't? Who gets to be paid more?

As anyone who's ever worked with a skilled administrator or community manager can tell you, "non-technical" roles require a deep level of expertise in highly sophisticated work. Their work *is* technical. But our industry's reliance on this distinction — on dividing "technical" roles from "non-technical" ones — primarily has value to those in power. It allows investors like the ones we've discussed — and the CEOs of the companies in which they invest — to telegraph how they assign value to different classes of people.

I'd like to suggest that there's a similar problem inherent in the term "tech worker" — that our existing definition is entirely too narrow, and rooted in incredibly classist assumptions about who does tech work. As we've discussed, there is a high and hidden cost to the creation of digital products and platforms, one foisted upon an entire class of workers that's been made largely invisible. They are paid hourly and poorly, often receive no benefits, and have nonexistent job security — but they are directly and literally responsible for making our products and platforms *function*. Why aren't they considered "tech workers" too?

After all, it's worth noting that contracted workers have borne the brunt of the industry's firings.

- In a much quieter round of firings at Twitter, an estimated five thousand contractors were summarily and suddenly laid off in November 2022 (https://ydatu.com/05-41).
- That same month, *Forbes* reported that as Meta continued to cull its corporate workforce, the company had begun cutting contracts with the companies that provide various on-site support, like janitorial services and drivers for employee shuttles (https://ydatu.com/05-42).

- In the second quarter of 2022 alone, Amazon laid off approximately *one hundred thousand* warehouse workers (https://ydatu.com/05-43).

Contrast those numbers with this excerpt from the *New York Times*, reporting the news that Amazon was laying off ten thousand corporate and technology jobs:

> Changing business models and the precarious economy have set off layoffs across the tech industry. Elon Musk halved Twitter's head count [by firing approximately 3,700 salaried workers] this month after buying the company, and last week, Meta, the parent company of Facebook and Instagram, announced it was laying off 11,000 employees, about 13 percent of its work force. (https://ydatu.com/05-44)

Those numbers are awful, truly. But they're also much, much lower than they ought to be, as they're exclusively reporting the layoffs of *salaried* workers. Read the paragraph again, and note how salaried corporate workers are cited as significant losses; hourly or contract workers aren't even mentioned, presumably because their precarity is seen as an acceptable cost of doing business in the tech industry. That it's perfectly acceptable — *normal,* even — to lay off thousands and thousands of low-paid, hourly workers. So the next time you read coverage of the layoffs from our industry's largest companies, pay attention to who's allowed to be characterized as a "worker" in the tech industry...and who isn't.

I'll say it again: during the wave of layoffs in 2022 and 2023, the overwhelming majority of people who lost their jobs in the tech industry weren't designing or building digital products. They were working in fulfillment warehouses, or in cafeteria kitchens; they were driving buses, or cleaning the hallways of corporate buildings. They were responding to customer emails, or sifting through traumatizing imagery and hateful speech as they sanitized our social media feeds. And they were doing all this work — this *tech work* — for exceedingly low pay, often without benefits, and without a guarantee that they would still have a job tomorrow.

Working in this industry requires holding two truths in your head at once: that the tech industry produces tremendous amounts of capital; and that most of its workers are toiling under atrocious working conditions, and living in financial precarity. However, these truths aren't contradictory: they're deeply, deeply linked. The digital products we design and build depend upon these vulnerable and exploited workers. These exploited tech workers are directly responsible for the success of our industry; without their labor, our industry's profits would dwindle. That's why the notion that "tech work" is done only by a privileged class of knowledge workers is a harmful idea—one we need to actively, aggressively dismantle.

THE WORLD WIDE WORK

Why does any of this matter?

Well, I'll start here: the sociologist and labor scholar Stanley Aronowitz once argued that over the last several decades, we've been watching a large, slow-moving trend play out across the world. Namely, that global economies have been gradually transforming the working class, pushing workers away from stable, salaried jobs, and toward work that is almost entirely transient, contracted, and temporary. As a result, there's a new class of worker defined almost entirely by a lack of stability—what Aronowitz calls *the precariat*:

> The rise of temporary work as a labor solution has cast a huge shadow over the professional and technical caste. People must take these jobs, because a paycheck is still the major source of food on the table and a roof over one's head. But very few jobs now offer real fulfillment, professional or material. For millions, life has become precarious. Increasingly, the job as an institution is under siege, because employers—public and private—hire only on a contingent or contract basis and do not offer health care coverage, pensions, or paid holidays and vacations to a substantial portion of their workforce. In short, the precariat has expanded to include the new middle class, and once privileged professionally and technically qualified workers have even joined

the proletariat, working in service jobs when they cannot find work — even precarious work — in their own fields. (https://ydatu.com/05-45)

We've looked at several examples in this chapter of the tech industry's precariat — data trainers, social media moderators, warehouse workers, and hourly contractors. But there are so many we haven't discussed: drivers for transportation services like Uber or Lyft; drivers for on-demand delivery services like Doordash, Uber Eats, or GrubHub; hourly employees at retail stores for tech giants like Apple, Google, or Amazon; and so many more. If we want to grapple with the injustices our industry is built upon, it's imperative we see these workers *as* tech workers. If we can manage that, we can fight alongside them to right those wrongs.

But in doing so, it's important to note that *every* tech worker is being pushed toward precarity. And yes, that includes more privileged tech workers like me — and perhaps you, too.

I want to be very clear: there are many axes of privilege at play here, and they can't be overlooked. Some of us perform tech work in considerable comfort and safety; many, many more tech workers labor in exploitative, dangerous conditions. We have to see and acknowledge that divide between us, if we hope to close it.

With that said, all of us — *all of us* — work in a system that is actively trying to devalue our work. The companies we work for are moving to automate our labor, and working to push us away from any semblance of long-term financial stability. But we have the power to change that. Not individually, but collectively — with our unions.

As my grandmother would've said, here's the way of it: our industry has engaged in various ethical and moral lapses in the pursuit of scale. And with that scale, there's a precarity visited upon every single tech worker. And as tech workers, we need to unionize to build power, which will allow us to protect our benefits, our workplaces, and ourselves.

But we can't stop there: we need to unionize because it's a path to power we can use for *all* tech workers in our industry, especially those who haven't been historically thought of as "tech workers." If we can build power *together*, across those lines, then something truly remarkable will happen. Unions have, at the best moments in their history, stood in solidarity with each other. There is no fight we can't win once we've unionized our industry's warehouses *and* its offices. It's time we do just that.

What does that look like? Well, we've seen one major example of that already, of the advocacy done during the 2018 Google walkouts, to fight for protections that covered salaried and contract workers alike. But in terms of what's next, well — I can't help but see glimpses of that future in the pre-majority unions at both Apple and Alphabet. At the time of this writing, workers at two Apple retail stores have unionized, with more campaigns underway — and those wins have been supported, at least in part, by Apple Together, the company's pre-majority union. And the Alphabet Workers Union has been fighting for the rights of the temporary and contract workers across the corporation, and helped Google Fiber employees form their first union in early 2022 — *with* bargaining rights (https://ydatu. com/05-46). Treating your premajority union as an incubator for contractors' majority unions? Pretty brilliant. What's more, it's a tangible example of solidarity between contracted and salaried tech workers.

And it's not the only one. In fact, one of the reasons the AWU opted to become a premajority union is *because* they wanted to include non-salaried temps and contractors, who wouldn't be eligible to join a majority union:

> **Our union** strives to protect Alphabet workers, our global society, and our world. We recognize our power as **Alphabet workers** — full-time employees, temporary employees, vendors, and contractors — comes from our solidarity with one another and our ability to collectively act to ensure that our workplace is equitable and Alphabet acts ethically. (https:// ydatu.com/05-47)

But as we look to build these connections, this *solidarity* across different groups of tech workers, we need to be mindful of more marginalized tech workers' history in this fight. Because frankly, many of them have been organizing for a long time:

1. The Teamsters have been organizing Bay Area shuttle bus drivers for years, helping contract drivers at Google unionize in 2016, and at Facebook in 2014 (https://ydatu.com/05-48, subscription required).
2. Since 2018, the union Unite Here has been unionizing thousands of contract cafeteria workers across the tech industry, most recently four thousand workers at Google (https://ydatu.com/05-49, subscription required).
3. In the first months of 2022, workers at Amazon's JFK8 fulfillment center in New York City made history, voting to form the first union at the ecommerce giant (https://ydatu.com/05-50, subscription required).

I mention this because these tech workers have so much to teach us — lessons from the fights they've won, the fights they've lost, and how they're planning to move forward.

But just as importantly, their history is a reminder that tech unions are not new. As I mentioned in the introduction to this book, it's tempting to think that unions were only recently introduced to the industry, but these tech workers — the tech workers toiling under the worst parts of our industry — have been unionizing for a long, long time. Unions have been in the modern tech industry for nearly a decade.

I'll say it again: *tech unions are not new*. There's already a rich history of labor organizing in our industry. We just have to learn from it.

Now, I'm not here to suggest that unions will cure all the industry's ills. The tech industry's issues are too vast, too complicated for there to be a *single* solution. To properly address them, we'll need a variety of approaches: regulatory fixes, organizational and cultural change, worker cooperatives, and more.

But unions are a critical part of the answer. When I interviewed the labor researcher Emily Mazo, she reminded me that unions aren't just a path to power in the workplace — they're a path to broader political and cultural power, too:

A union is also important for winning social justice fights outside of the workplace, nationally and internationally. But we can only do that if we all organize, and unions are the place to do that organizing. When we all organize together to build labor power, we can then use that labor power: to fight for the things that we want to win.

The tech industry's labor movement is giving us something incredibly precious: *time*. Unions — and through them, the contracts we'll win — will afford us the protections we need, the stability we need, in order to take a moment to imagine better futures for ourselves and our industry. That's not to suggest that those contract wins aren't worthwhile, or that we shouldn't fight like hell for them — they are, and we should. But we need time, too: time to dream, time to organize, time to define a better form of tech work. If we can win our unions, then we can *keep* fighting for a better tech industry — one that's truly worth working in.

I'm glad to be here, fighting by your side. Let's get to work.

ACKNOWLEDGMENTS

THANK YOU FOR READING this little book. It means the world to me.

This book owes a heavy debt to the workers, activists, organizers, and scholars I interviewed. Over the course of a year and a half I reached out to nearly fifty people, many of them strangers, and I was moved time and again by their willingness to share their stories with me. Many of them spoke with me on the record; some were only able to speak confidentially. I took each and every interview as a sign of trust in this book and its author, and I dearly hope I did their stories justice. If any of you happen to read this: thank you so much for your time, and for the work you do in making our industry better for its workers.

I'd like to extend a special note of thanks to Clarissa Redwine, RV Dougherty, Keith Hogarty, Nozlee Samadzadeh, Kathy Zhang, Jenny Mackintosh, Afton Cyrus, and Chelsea Noriega. Each of them took the time to connect me with workers and organizers in their network, often vouching for me directly.

As it happens, RV Dougherty was one of my very first interviews for this book. I'd reached out to them because their work at OPEIU's Tech Local 1010 is a constant inspiration to me, and I'm incredibly honored they agreed to act as this book's subject matter reviewer. I can't begin to tell you the many ways this book is better for their thoughtful feedback.

I'd like to thank Katel LeDû and the entire team at A Book Apart for originally publishing this book. In particular, I'd like to thank Caren Litherland for her keen eye and her careful edits; this book has deeply benefited from her work. And I'm pretty sure he knows this, but I'm deeply grateful to Jason Santa Maria, who provided near-endless support, feedback, and encouragement as I reacquired the rights to this book.

And of course, a special note of thanks goes to Lisa Maria Marquis, who first approached me about writing this book, and then shepherded it all the way to its first draft. If it weren't for her sharp, compassionate edits, her patient encouragement, and her seemingly endless support, this book would be a pile of half-baked ideas and ellipses. Thank you, Lisa Maria.

By now, you've probably figured out that Ursula Franklin has been an inspiration to me in so many ways; I wish I could have had the opportunity to thank her for everything she taught me. But I *can* thank my friends Mandy Brown, Ingrid Burrington, and Deb Chachra, each of whom led me to Franklin's work in their own way.

A heartfelt shout-out to Erin Kissane, who kindly and enthusiastically looked over a draft of this book, and gifted me with a trove of wildly helpful feedback. Erin, I'm so glad to know you.

Siena Chiang and Roy Bahat have both been incredibly kind toward this book, and toward me. They offered support, encouragement, and introductions, and I'm grateful to them both.

My thanks to Kate Bahn and Kathryn Zickuhr, from the Washington Center for Equitable Growth. Their advice and perspective were a massive help, as were the many resources and articles they graciously sent my way.

The first union poster I ever saw was on the wall of Garret Keizer's library, which feels fitting: much of what I know about unions, about labor, and about solidarity with your fellows, I owe to him. Heck—I owe much of what I know about writing to him. He gave me endless amounts of support and encouragement as I wrote this little book—and throughout the decades that led up to it. I hope I made you proud, Garret.

My wife Elizabeth Galle made this project possible in more ways than I'll ever be able to count. This book, and everything else I do, is for her.

RESOURCES

I AM A RELATIVELY NEW student to labor history, and my education has come from the writing and research of countless historians, journalists, organizers, activists, and workers. If I have one wish for this book, it's that it will act as an introduction to the voices that taught me.

In this section, I've included a few key pieces that guided me through this book. I hope they lead you down some interesting paths.

Books on labor, its history, and the law

I kept five books close at hand while I wrote:

- *A Collective Bargain: Unions, Organizing, and the Fight for Democracy* by Jane McAlevey (https://ydatu.com/06-01)
- *Labor Law for the Rank and Filer: Building Solidarity While Staying Clear of the Law* by Staughton Lynd and Daniel Gross (https://ydatu.com/06-02)
- *Fight Like Hell: The Untold History of American Labor* by Kim Kelly (https://ydatu.com/06-03)
- *A History of America in Ten Strikes* by Erik Loomis (https://ydatu.com/03-02)
- *Subterranean Fire: A History of Working-Class Radicalism in the United States* by Sharon Smith (https://ydatu.com/06-04)

I'd like to note that the last three books were invaluable sources for the histories I've covered in this book. But I owe a debt to all five of these titles, and to their authors. (Of course, any errors that remain are mine alone.)

Online resources

- *The Voice of Industry* was a worker-run newspaper published by young American women, filled with coverage and criticism of the labor issues rampant in the 1840s. Thankfully, its issues have been digitized and archived online (https://ydatu.com/06-05).

- Collective Action in Tech is a volunteer-run organization dedicated to covering worker-led actions in the tech industry, and to publishing profiles of the tech organizers fighting for a better future (https://ydatu.com/06-06).
- Interintellect is an online community that hosts countless online events, and their (free!) series of talks on labor and organizing has been absolutely stellar (https://ydatu.com/06-07).
- The Zinn Education Project offers a trove of *excellent* labor-related teaching materials, histories, and resources (https://ydatu.com/06-08).
- The Kickstarter United Oral History is a podcast that charts the formation of Kickstarter's groundbreaking union, "from genesis all the way to the vote win." (https://ydatu.com/06-09)

Organizing guides and training

- I'd strongly recommend *Secrets of a Successful Organizer*, a short guide to effective organizing published by the team at Labor Notes (https://ydatu.com/06-10).
- Bill Fletcher's *"They're Bankrupting Us!": And 20 Other Myths About Unions* directly addresses many of the myths surrounding unions, and might be helpful to you in your organizing conversations with coworkers — and perhaps again during a management fight (https://ydatu.com/06-11).
- Labor Notes is a nonprofit that provides union members and labor activists with the training and community they need. Toward that end, the nonprofit offers free online workshops, as well as an annual conference (https://ydatu.com/06-12).
- The Emergency Workplace Organizing Committee is a grassroots organizing program designed to support workplace organizing with resources and free online training (https://ydatu.com/06-13).
- Collective Action in Tech published *DMs Open*, a free zine about winning your remote organizing campaign (https://ydatu.com/06-14).

Staying informed on labor issues

It's never been easier to stay abreast of the latest labor news and developments. Here are some of my favorite ways to do just that.

- In addition to everything else they do, Labor Notes also publishes a monthly magazine covering labor issues and union fights (https://ydatu.com/06-15).
- CWA's CODE-CWA and OPEIU's Tech Local 1010 both have free newsletters I recommend, in which they update readers on the respective unions' activities. You can subscribe at https://ydatu.com/06-16 and https://ydatu.com/04-06, respectively.
- Lauren Kaori Gurley (https://ydatu.com/06-17) and Caroline O'Donovan (https://ydatu.com/06-18) are two reporters whose coverage of the modern labor movement has been endlessly helpful to me.
- Published by the United Electrical, Radio and Machine Workers of America (UE), *UE NEWS* is a print periodical containing in-depth coverage of labor news and issues (https://ydatu.com/06-19).
- *In These Times* is an American nonprofit news organization offering coverage of the labor beat. It is excellent, and very much worth your time (https://ydatu.com/06-20).
- Before her passing, Jane McAlevey wrote a magnificent labor column for *The Nation*. As with everything McAlevey wrote, it's filled with keen insight, clear-eyed strategies, and timely lessons (https://ydatu.com/06-21).
- Economic Policy Institute is a nonprofit think tank that aims to center low- and middle-income workers in policy decisions; its newsletter is excellent (https://ydatu.com/06-22).
- Washington Center for Equitable Growth is a nonprofit research organization that studies economic inequality in the United States, and labor is one of their key focuses. As it happens, they also offer a free newsletter (https://ydatu.com/06-23).
- Tim Carmody's *Amazon Chronicles* is a newsletter covering the planet's largest retailer, digging into its plans, its ambitions — and its labor issues (https://ydatu.com/06-24).

Talks and interviews

- Vicki Crosson and Shane Moore, two members of the NYT Tech Guild, gave a talk at the 2022 Strange Loop conference on the technology and tools their union used to manage its organizing (https://ydatu.com/06-25, video).
- In 2019, Jane McAlevey gave a wide-ranging interview with *Current Affairs*, in which she discussed organizing strategies and finding electoral and political power through unions, and expanded on her distinction between "mobilizing" and "organizing" (https://ydatu.com/06-26).
- Labor Notes hosted an event in March 2023 discussing LGBTQIA+ labor histories, and invited speakers from Pride at Work Boston, Howard Brown Health, and the Alphabet Workers Union. Videos and resources from the event have been posted online (https://ydatu.com/06-27).

Workplace surveillance and union-busting

- This research paper from 2022 looks at the rise of labor organizing in tech workplaces, and studies the privacy habits adopted by organizers to mitigate the risk of retaliation (https://ydatu.com/06-28).
- The *Intercept* looks at a recent trend among union-busting law firms, and how they've embraced the language of diversity and inclusion to mitigate worker organizing (https://ydatu.com/06-29).
- *Protocol* reported on the union-busting campaign that defeated the Mapbox union election, and tied it to a larger wave of anti-union mobilization sweeping across tech companies (https://ydatu.com/04-20).
- In an essay for *Boston Review*, Brishen Rogers looked at technology as a tool for corporate surveillance in the workplace, and argued for greater worker control in response to it (https://ydatu.com/06-30).
- A 2021 research paper by Kathryn Zickuhr outlined how worker monitoring and surveillance has become the norm in modern workplaces, and suggested several structural and regulatory fixes for addressing the harms caused by these practices (https://ydatu.com/06-31).

- *Vox* published an in-depth look at a leaked memo detailing Amazon's strategy for combating unions: namely, a massive, tightly orchestrated public relations campaign (https://ydatu.com/06-32).

On marginalized tech workers

- I was introduced to the concept of underpaid data trainers by this 2018 BBC feature profiling Brenda Monyangi, a single mother who works at a data annotation company in Nairobi (https://ydatu.com/06-33).
- I first learned about Google's Scan Ops workers in "The Invisible Labor of the Web", Sarah Adams's talk at the 2019 Theorizing the Web conference (https://ydatu.com/06-34). It's a marvelous talk, and I still think about it.
- *Ghost Work*, by Mary L. Gray and Siddharth Suri, is a searing look at the invisible labor powering the modern tech industry (https://ydatu.com/06-35).
- During the tech layoffs at the end of 2022, both *Time* (https://ydatu.com/06-36) and Bloomberg (https://ydatu.com/06-37) reported on the outsize impact the layoffs had on H1-B visa holders.
- The *Markup* published a data-driven look at the violence faced by gig economy drivers and couriers (https://ydatu.com/06-38). *Please note*: this article contains graphic descriptions of violence.
- Adrienne Williams, Milagros Miceli, and Timnit Gebru wrote a remarkable article on the exploited labor behind the modern "artificial intelligence" movement (https://ydatu.com/06-39).
- Mandy Brown's "Smoke Screen" is an incisive, thoughtful essay on "artificial intelligence" and the value of human labor (https://ydatu.com/06-40).

REFERENCES

Shortened URLs are numbered sequentially; the related long URLs are listed below for reference.

Chapter 1

01-01 https://wwhpchicago.org/the-reverend-addie-l.-wyatt.html
01-02 https://bookshop.org/p/books/the-real-world-of-technology-ursula-franklin/9995479
01-03 https://data.census.gov/table?tid=ACSST1Y2021.S1901
01-04 https://www.wsj.com/articles/what-alphabet-meta-and-other-s-p-500-firms-paid-workers-last-year-11654084801
01-05 https://www.protocol.com/bulletins/meta-laundry-leftovers
01-06 https://www.nytimes.com/2022/02/07/business/amazon-salary-cash-cap.html
01-07 https://www.theverge.com/2022/7/12/23206113/google-ceo-sundar-pichai-memo-hiring-slowdown-2022
01-08 https://layoffs.fyi/
01-09 https://www.computerworld.com/article/3682071/amazon-layoffs-now-expected-to-mount-to-20000-including-top-managers.html
01-10 https://about.fb.com/news/2022/11/mark-zuckerberg-layoff-message-to-employees/
01-11 https://www.nytimes.com/2023/03/14/technology/meta-facebook-layoffs.html
01-12 https://www.theverge.com/2023/1/18/23560315/microsoft-job-cuts-layoffs-2023-tech
01-13 https://www.nytimes.com/2023/01/20/business/google-alphabet-layoffs.html
01-14 https://stackoverflow.blog/2023/04/02/the-people-most-affected-by-the-tech-layoffs/
01-15 https://projectinclude.org/remote-work-report/
01-16 https://hired.com/wage-inequality-report/2021/
01-17 https://www.theverge.com/23551060/elon-musk-twitter-takeover-layoffs-workplace-salute-emoji
01-18 https://twitter.com/esthercrawford/status/1587709705488830464
01-19 https://www.bbc.com/news/technology-63897608
01-20 https://stripe.com/en-au/newsroom/news/ceo-patrick-collisons-email-to-stripe-employees

Chapter 2

02-01 https://bookshop.org/p/books/what-we-build-with-power-the-fight-for-economic-justice-in-tech-david-delmar-senties/18572078

02-02 https://www.ohchr.org/Documents/HRBodies/HRCouncil/FFM-Myanmar/A_HRC_39_64.pdf

02-03 https://aws.amazon.com/rekognition/

02-04 https://www.theguardian.com/commentisfree/2021/may/18/amazon-ring-largest-civilian-surveillance-network-us

02-05 https://www.googlecloudpresscorner.com/2022-02-09-Defense-Innovation-Unit-Selects-Google-Cloud-for-Secure-Cloud-Management-Implementation

02-06 https://www.bloomberg.com/news/articles/2022-09-01/microsoft-combat-goggles-win-first-us-army-approval-for-delivery

02-07 https://twitter.com/USArmy/status/1384538038571880448

02-08 https://www.accessible-archives.com/collections/godeys-ladys-book/

02-09 https://digital.hagley.org/Singer_S81_4#page/2/mode/2up

02-10 https://www.flickr.com/photos/kheelcenter/5279597520/in/photostream/

02-11 https://webfoundation.org/faq/#2

02-12 https://webfoundation.org/

02-13 https://neveragain.tech/

02-14 https://www.rollingstone.com/politics/politics-news/this-is-the-prison-like-border-facility-holding-migrant-children-628728/

02-15 https://github.blog/2019-10-09-github-and-us-government-developers/

02-16 https://www.washingtonpost.com/context/letter-from-github-employees-to-ceo-about-the-company-s-ice-contract/fb280de9-2bc3-40d5-b1a5-e3b954bf0d25/

02-17 https://amazonemployees4climatejustice.medium.com/amazon-employees-are-joining-the-global-climate-walkout-9-20-9bfa4cbb1ce3

02-18 https://nytimes.github.io/nyt-solidarity/

02-19 https://www.linkedin.com/posts/timothy-j-aveni_blacklivesmatter-activity-6673316720993824768-q_dU/

02-20 https://www.nytimes.com/2018/05/30/technology/google-project-maven-pentagon.html

02-21 https://www.theverge.com/2018/11/27/18114285/google-employee-china-censorship-protest-project-dragonfly-search-engine-letter

02-22 https://www.nytimes.com/2018/10/25/technology/google-sexual-harassment-andy-rubin.html

02-23 https://mobile.twitter.com/search?q=from%3AGoogleWalkout&src=typed_query&f=image

02-24 https://www.businessinsider.com/google-walkout-live-pictures-of-protesting-google-workers-2018-11

02-25 https://googlewalkout.medium.com/google-employees-and-contractors-participate-in-global-walkout-for-real-change-389c65517843

02-26 https://www.dol.gov/general/aboutdol/hallofhonor/1990_debs

02-27 https://www.thecut.com/2018/11/google-walkout-organizers-explain-demands.html

02-28 https://www.elle.com/culture/tech/a30259355/google-walkout-organizer-claire-stapleton

02-29 https://www.wired.com/story/tech-organizing-labor-ibm-history/

02-30 https://data.collectiveaction.tech/?query=protest,white%20collar%20workers

02-31 https://data.collectiveaction.tech/

02-32 https://eli.naeher.name/pdfs/interrupt-14.pdf

02-33 https://eli.naeher.name/computer-people-for-peace/

02-34 https://theoutline.com/post/4029/computer-people-for-peace-history

02-35 https://news.techworkerscoalition.org/

02-36 https://news.techworkerscoalition.org/2021/07/06/issue-15/

02-37 https://news.techworkerscoalition.org/2022/07/05/issue-11/

02-38 https://jacobin.com/2018/06/google-project-maven-military-tech-workers

02-39 https://kickstarterunited.org/

02-40 https://abtwu.org/

Chapter 3

03-01 https://www.google.com/books/edition/American_Illustrated_Magazine/LGhEAQAAMAAJ?hl=en&gbpv=0

03-02 https://thenewpress.com/books/history-of-america-ten-strikes

03-03 https://www.industrialrevolution.org/10-hours-movement.html

03-04 https://bookshop.org/p/books/fight-like-hell-the-untold-history-of-american-labor-kim-kelly/18018543

03-05 https://www.facinghistory.org/sites/default/files/2022-07/Petition_from_the_Colored_Washerwomen.pdf

03-06 https://nvdatabase.swarthmore.edu/content/lawrence-ma-factory-workers-strike-bread-and-roses-us-1912

03-07 https://www.zinnedproject.org/materials/bread-and-roses-strike-story/

03-08 https://www.nlrb.gov/

03-09 https://itsgoingdown.org/c19-mutual-aid/

03-10 https://kickstarterunited.org/ratified_CBA_KSRU.pdf

03-11 https://kickstarterunited.org/first-contract/

03-12 https://cfaworkersunited.com/

Chapter 4

04-01 https://www.versobooks.com/books/3038-old-gods-new-enigmas

04-02 https://nytimesguild.org/tech/

04-03 https://www.amazonlaborunion.org/

04-04 https://cwa-union.org/

04-05 https://www.theverge.com/2021/3/2/22307671/glitch-workers-sign-historic-collective-bargaining-agreement-cwa

04-06 https://www.opeiu.org/

04-07 https://www.techworkersunion-1010.org/

04-08 https://www.nlrb.gov/about-nlrb/rights-we-protect/the-law/jurisdictional-standards

04-09 https://www.ueunion.org/ue-news-feature/2022/seventy-five-years-later-toll-of-taft-harley-weighs-heavily-on-labor

04-10 https://www.gao.gov/products/gao-21-242

04-11 https://www.vox.com/podcasts/2020/3/17/21182149/jane-mcalevey-the-ezra-klein-show-labor-organizing

04-12 https://nytimesguild.org/papers/2022-02-07-solidarity-with-salary-sharing/

04-13 https://catalist.us/catalist-recognizes-staff-union/

04-14 https://cwa-union.org/news/releases/tech-workers-app-developer-glitch-vote-form-union-and-join-cwa-organizing-initiative

04-15 https://www.epi.org/publication/union-avoidance/?mc_cid=2c35837598

04-16 https://arstechnica.com/tech-policy/2022/01/google-hired-union-busting-consultants-to-convince-employees-unions-suck/

04-17 https://www.vice.com/en/article/3a8vyk/kickstarter-hired-a-law-firm-that-advertises-maintaining-a-union-free-workplace

04-18 https://www.theverge.com/2022/4/25/23041632/apple-hires-anti-union-lawyers-littler-mendelson-union-fight-cwa

04-19 https://www.protocol.com/workplace/why-tech-companies-union-bust

04-20 https://www.nlrb.gov/about-nlrb/rights-we-protect/your-rights/weingarten-rights

04-21 https://laborcenter.berkeley.edu/turning-the-tables-participation-and-power-in-negotiations

Chapter 5

05-01 https://airbnb.design/sketching-interfaces/

05-02 https://commons.wikimedia.org/wiki/File:A_mechanical_dove_8274822e-52fe-40fa-ac4d-f3cde5a332ae.png

05-03 https://youtu.be/outcGtbnMuQ?t=976

05-04 https://daverupert.com/2022/08/github-copilot/

05-05 https://twitter.com/JasonColavito/status/1611710986871767041

05-06 https://www.theverge.com/2023/3/30/23663206/buzzfeed-ai-travel-guides-buzzy

05-07 https://www.theverge.com/2023/1/19/23562966/cnet-ai-written-stories-red-ventures-seo-marketing

05-08 https://www.wgacontract2023.org/the-campaign/wga-negotiations-status-as-of-5-1-2023

05-09 https://www.bloomberg.com/news/articles/2022-10-17/digital-media-firm-stability-ai-raises-funds-at-1-billion-value

05-10 https://observer.com/2023/01/chatgpt-openai-valued-29-billion/

05-11 https://twitter.com/pmarca/status/1515195878604087297

05-12 https://nat.org/

05-13 https://medium.com/@alt.cap/time-to-get-fit-an-open-letter-from-altimeter-to-mark-zuckerberg-and-the-meta-board-of-392d94e80a18

05-14 https://www.tcifund.com/files/corporateengageement/alphabet/15th%20November%202022.pdf

05-15 https://www.tcifund.com/files/corporateengageement/alphabet/20th%20January%202023.pdf

05-16 https://www.theguardian.com/business/2021/mar/05/the-very-private-life-of-the-man-on-britains-biggest-salary

05-17 https://www.nytimes.com/2023/02/28/technology/twitter-outages-elon-musk.html

05-18 https://www.vox.com/technology/2023/3/23/23651151/twitter-advertisers-elon-musk-brands-revenue-fleeing

05-19 https://www.bloomberg.com/news/articles/2023-01-23/microsoft-makes-multibillion-dollar-investment-in-openai

05-20 https://www.foundsf.org/index.php?title=Mechanization_on_the_Waterfront

05-21 https://laborcenter.berkeley.edu/wp-content/uploads/2020/12/Working-Paper-Union-Collective-Bargaining-Agreement-Strategies-in-Response-to-Technology.pdf

05-22 https://theartofgooglebooks.tumblr.com

05-23 https://theartofgooglebooks.tumblr.com/tagged/hand

05-24 https://theartofgooglebooks.tumblr.com/post/692755692798558208/holding-down-a-torn-page-from-p-1124-25-of

05-25 https://about.google/belonging/diversity-annual-report/2022/

05-26 http://www.andrewnormanwilson.com/ScanOps.html

05-27 https://vimeo.com/15852288

05-28 https://time.com/6247678/openai-chatgpt-kenya-workers/

05-29 https://www.wired.com/2014/10/content-moderation/

05-30 https://www.theverge.com/2019/2/25/18229714/cognizant-facebook-content-moderator-interviews-trauma-working-conditions-arizona

05-31 https://www.businessinsider.com/facebook-median-pay-240000-2017-2018-4

05-32 https://www.amazon.science/latest-news/how-amazon-robotics-is-working-on-new-ways-to-eliminate-the-need-for-barcodes

05-33 https://amazonchronicles.ghost.io/amazons-heroic-phase
05-34 https://www.nytimes.com/interactive/2021/06/15/us/amazon-workers.html
05-35 https://www.thecourier.co.uk/fp/news/fife/325800/exclusive-amazon-workers-sleeping-in-tents-near-dunfermline-site/
05-36 https://revealnews.org/article/the-truth-about-injuries-at-amazon/
05-37 https://www.ehstoday.com/safety/article/21919294/two-worker-deaths-in-september-at-different-amazon-warehouses-spawn-concern-from-worker-advocates
05-38 https://www.theatlantic.com/technology/archive/2019/11/amazon-warehouse-reports-show-worker-injuries/602530/
05-39 https://www.thedailybeast.com/amazon-employee-who-died-on-prime-day-rafael-reynaldo-mota-frias-was-hardworking-dad
05-40 https://everythingchanges.us/blog/no-one-is-non-technical/
05-41 https://www.theverge.com/2022/11/13/23456554/twitter-reportedly-cut-thousands-contractors-without-warning-layoffs-elon-musk
05-42 https://www.forbes.com/sites/cyrusfarivar/2022/11/22/meta-layoffs-contract-workers/
05-43 https://www.barrons.com/articles/amazon-jobs-cuts-staff-51659122765
05-44 https://www.nytimes.com/2022/11/14/technology/amazon-layoffs.html
05-45 https://www.versobooks.com/books/1983-the-death-and-life-of-american-labor
05-46 https://www.theverge.com/2022/3/25/22996053/google-fiber-union-contractors-workers-vote-cwa
05-47 https://alphabetworkersunion.org/principles/mission-statement/
05-48 https://news.bloomberglaw.com/daily-labor-report/google-shuttle-drivers-gain-teamsters-representation
05-49 https://www.washingtonpost.com/technology/2022/09/05/google-union-pandemic/
05-50 https://www.nytimes.com/2022/04/01/technology/amazon-union-staten-island.html

Resources

06-01 https://bookshop.org/p/books/a-collective-bargain-unions-organizing-and-the-fight-for-democracy-jane-mcalevey/16304977
06-02 https://pmpress.org/index.php?l=product_detail&p=325
06-03 https://rep.club/products/fight-like-hell
06-04 https://www.haymarketbooks.org/books/1166-subterranean-fire-updated-edition
06-05 https://industrialrevolution.org/
06-06 https://collectiveaction.tech/about/
06-07 https://interintellect.com/series/book-talks-learning-about-labor/
06-08 https://www.zinnedproject.org/themes/labor/

06-09 https://kickstarterunited.org/oral-history/
06-10 https://www.labornotes.org/secrets
06-11 https://billfletcherjr.com/books/theyre-bankrupting-us/
06-12 https://labornotes.org/events
06-13 https://workerorganizing.org/training/
06-14 https://collectiveaction.tech/2022/dms-open/
06-15 https://www.labornotes.org/archives
06-16 https://code-cwa.org/
06-17 https://twitter.com/LaurenKGurley
06-18 https://twitter.com/ceodonovan/
06-19 https://www.ueunion.org/ue-news
06-20 https://inthesetimes.com/labor
06-21 https://www.thenation.com/authors/jane-mcalevey/
06-22 https://www.epi.org/signup/
06-23 https://equitablegrowth.org/engage/get-updates/
06-24 https://amazonchronicles.ghost.io/
06-25 https://www.youtube.com/watch?v=-clXJvIjP84
06-26 https://www.currentaffairs.org/2019/04/jane-mcalevey-on-how-to-
       organize-for-power
06-27 https://labornotes.org/blogs/2023/03/videos-mighty-gay-unions-
       queer-and-trans-labor-histories-and-futures
06-28 https://arxiv.org/abs/2206.00035
06-29 https://theintercept.com/2022/06/07/union-busting-tac-
       tics-diversity/
06-30 https://www.bostonreview.net/articles/workplace-data-is-a-tool-
       of-class-warfare/
06-31 https://equitablegrowth.org/research-paper/workplace-surveil-
       lance-is-becoming-the-new-normal-for-u-s-workers/
06-32 https://www.vox.com/recode/23282640/leaked-internal-memo-re-
       veals-amazons-anti-union-strategies-teamsters
06-33 https://www.bbc.com/news/technology-46055595
06-34 https://docs.google.com/presentation/d/1y5jLzfa9rpGyP-gZHy-
       wCS7Lxrn6wucwVTF3o_6khIVE/edit#slide=id.g566f89cb4e_2_52
06-35 https://rep.club/products/ghost-work
06-36 https://time.com/6239846/tech-layoffs-visa-h1b/
06-37 https://www.bloomberg.com/news/articles/2022-11-21/2022-tech-
       layoffs-leave-h-1b-visa-holders-in-limbo
06-38 https://themarkup.org/working-for-an-algorithm/2022/07/28/
       more-than-350-gig-workers-carjacked-28-killed-over-the-
       last-five-years
06-39 https://www.noemamag.com/the-exploited-labor-behind-artifi-
       cial-intelligence/
06-40 https://aworkinglibrary.com/writing/smoke-screen

INDEX

ABOUT THE AUTHOR

 **Ethan Marcotte** is a web designer, speaker, and author. He's perhaps best known for creating responsive web design, which helped define a new way of designing for the ever-changing web. He's passionate about creating a web that's beautiful for everyone, everywhere.

Over his career, Ethan's clientele has included *New York Magazine*, the Sundance Film Festival, *The Boston Globe*, and *People Magazine*. Additionally, Ethan is a featured speaker at conferences across the globe, including Adobe MAX, SXSW Interactive, and Webstock.

COLOPHON

The text is set in Tiempos Text and Untitled Sans, both by Klim Type Foundry. The headlines and book cover use Cambon Condensed by General Type Studio, as well as Klim Type Foundry's Söhne.

The rose illustration used on the cover was originally sourced from freepik.com.

www.ingramcontent.com/pod-product-compliance
Lightning Source LLC
LaVergne TN
LVHW072122060326
832903LV00064B/4825